Position Location Techniques and Applications

Position Location Techniques and Applications

David Munoz
Frantz Bouchereau
Cesar Vargas
Rogerio Enriquez

ELSEVIER

AMSTERDAM • BOSTON • HEIDELBERG • LONDON
NEW YORK • OXFORD • PARIS • SAN DIEGO
SAN FRANCISCO • SINGAPORE • SYDNEY • TOKYO

Academic Press is an imprint of Elsevier

Academic Press is an imprint of Elsevier
30 Corporate Drive, Suite 400
Burlington, MA 01803

This book is printed on acid-free paper. ∞

Library of Congress Cataloging-in-Publication Data
Application submitted.

ISBN 13: 978-0-12-374353-4

For information on all Academic Press publications
visit our Website at *www.books.elsevier.com*

Printed in the United States
Transferred to Digital Printing, 2011

Contents

Preface

Purpose of This Book

Obtaining information on position location is, along with current regulatory requirements, one of the critical elements for the creation of location-based and context-aware services. It is also an important support for resource management and for the swift deployment of assistance services such as law enforcement.

The location problem has existed for many years and has motivated a great amount of research in and development of cellular-aided positioning algorithms and systems. In the satellite context, two global examples of such systems is the global positioning system (GPS) in the United States and the GALILEO system in the European Union. Diverse types of wireless network infrastructures are being deployed throughout residential areas, city commercial centers, university and company campuses, hospitals, amusement parks, and restaurants.

Wireless networking devices constitute the main infrastructure to be utilized for wireless location algorithms. Besides emergency services, several applications can be envisioned with these position estimation schemes. They include real-time display of self-location information; monitoring and mapping of potentially hazardous zones in disaster areas; fleet management and real-time traffic information retrieval; real-time amusement park, museum, and tourist guides; mobile databases; location-aware gaming systems; and location-sensitive billing.

Currently, several sources need to be consulted to obtain a comprehensive understanding of the location estimation problem, making the learning process lengthy and without a unified focus. The proliferation of mobile computing devices and wireless technologies has fostered a growing interest in the development of location-aware systems and services. Location estimation systems depend on a number of variables and scenarios that must be thoroughly understood, all at once, in order to cope with system design trade-offs, complexity issues, and design of efficient positioning algorithms.

However, most of the literature on cellular, ad hoc, and wireless sensor networks fails to treat the localization problem thoroughly. Very few books address the topic explicitly and, when they do, they tend to tackle it from the perspective of a single technology. Moreover, in most cases theoretical background and fundamental limits to the localization problem are neglected. For these reasons with this book, it is our intention to provide readers with the necessary location estimation expertise without their having to endure a cumbersome learning process. Further, we want to introduce the challenges and problems posed by the ever growing need to localize nodes in ad hoc and sensor networks.

The purpose of this book, then, is to develop a comprehensive and unified view of the wireless location-information acquisition problem and its solutions. More than providing an exhaustive survey of current location estimation techniques, we set a theoretical path that will allow practical engineers and researchers to understand and improve existing location schemes as well as to develop new ones within different scenarios.

The audience for this book includes those who work in, and end users of, related areas within the communications sector, such as operations, manufacturing, and service provision, as well as those in development engineering, consulting, academia, higher education, planning and resource management, and research. Applications-oriented professionals will also find support through discussions of the networking aspect of the problem, such as network architectures that support position location applications and services, as well as through discussion of the problem's algorithmic aspects.

Features

The presentation of the information in the book is self-contained, allowing readers to acquire the fundamentals for development, research, and operation of position location techniques. The book presents ideas and techniques such that readers will be capable of applying, evaluating, and extending them. Further, it introduces the ideas behind the algorithms, supplying concepts that show the generality of such techniques in terrestrial and satellite systems. The book can be considered as a reference by many engineers in the communications area and as a textbook in universities. It will also present a solid base for members of academia who are looking for an introduction to position location.

Although many books on position location can be found, most of them are technology driven or business oriented, which limits their scope to killer applications or services with a limited treatment of or orientation toward general scenarios that can be applied and extended to present and future scenarios and technologies. Other books concentrate on a particular technology, protocol, or scenario, limiting their usefulness once new technologies appear.

Information on location acquisition is sparse in the cellular systems literature. Few books address the topic explicitly, and those that do tend to tackle the problem from the perspective of a single technology with little emphasis on theory and fundamental limits. We provide a global, unified treatment of the position acquisition problem.

Trends in reconfigurable and multihop networks, which are not treated at all or only incompletely in other books, pose new challenges in location acquisition. For that reason, we revisit some traditional methodologies from a new perspective. We also consider some heuristic techniques suitable for a limited number of landmark references, multihop scenarios, and three-dimensional cases.

Organization

The material in this book is organized as follows. Chapter 1 addresses the relevance of position location (PL) information as a critical resource in improved network planning, development of new location-based services, fast deployment of assistance services as well as surveillance and security support, and many other applications such as fleet and crew management, environmental monitoring, and control. Conflicting criteria present in algorithm design due to PL information requirements, and achievable goals (e.g., accuracy, cost, processing load, implementation times) are reviewed in terms of service, environment, and technology.

As Chapter 1 discusses, location estimation methods can be implemented based on field intensity, angular or time-related measurements, or combinations of these factors. These signal measurements may be used to determine the length (range) or direction (angle) of the radio paths to/from a node of interest from/to multiple reference nodes. Alternatively, they may be used to constrain the position of a target node to circular or elliptical loci around multiple reference nodes.

Chapter 2 describes radio signal strength (RSS), angle of arrival (AOA), and time of arrival (TOA) measurements, and discusses techniques to estimate these variables, paying careful attention to the major sources of estimation errors such as channel fading, non-line of sight, and noise. Bounds on estimation performance of RSS, AOA, TOA, and range are also provided. A set of measurements originating at or being received by several reference nodes can be combined in optimal or suboptimal ways to estimate the position of a target node.

Chapter 3 presents optimal and suboptimal methods that combine RSS, AOA, TOA, and range measurements originating at several reference nodes to obtain estimates of the position of a target node. Cooperative and noncooperative data fusion techniques are discussed. Positioning algorithms are presented for static and dynamic scenarios. In the latter case, velocity is included in the model and filtering techniques are applied to predict the next position of the transmitting terminal. Introduced are important performance metrics that capture the system–network performance dependence, such as the dilution of precision that describes the amount by which estimation errors are degraded by the network geometry, and the Cramer-Rao bounds, which are lower bounds on the covariance of estimates of unbiased estimators.

In many scenarios, observable parameters are insufficient to provide accurate location information; however, broad location estimation may be feasible and sufficient for some purposes. In Chapter 4, heuristics applicable to both single-hop and multihop scenarios are introduced along with some of their major limitations. Sensitivity to errors or limited knowledge of the environment is also discussed, and practical and theoretical limitations such as measuring errors, impact of a reduced number of landmarks, multihop, uncertain node location, and mobility are presented.

In Chapter 5, we review the fundamentals of wireless networks to develop a thorough understanding of the basic functions involved in position location scenarios. PL is seen as part of the wireless network system; mobility is brought into the discussion because it imposes limits on network performance. It is also important to introduce how these scenarios evolve according to technology within new paradigms and architectures such as cooperative networks, cognitive radio, multihop scenarios, and reconfigurable networks.

In Chapter 6, the PL problem is treated as part of a network, with discussions and presentations of architectures and system points of view in different scenarios. Some technologies, such as WiFi, and ZigBee, are addressed as part of the general scenarios that have been presented up to this point. Sensor and ad hoc networks are also presented. Trade-offs that further comparison of different technologies are fundamental in PL decision making. This discussion includes received signal strength, time of arrival, and angle of arrival, as well as new ideas on connectivity and reachability in reconfigurable networks.

Chapter 7 describes the concepts behind the technology used by any satellite system for space positioning. The intention is twofold: a comprehensive development of underlying technological and scientific ideas and a foundation for appropriate and novel business applications. The basics of augmented systems are also presented.

Acknowledgments

The fundamental idea of a book on position location came from our research, course development at the graduate level, and publications, which would not have been possible without the support of the Nortel Networks–ITESM Monterrey Research Chair at the Center for Electronics and Telecommunications. We also must acknowledge the continuing support of the Mobility and Wireless Systems Research Chair at ITESM Monterrey and the many colleagues and research assistants who make up this group and who have contributed to our research and results over the years.

This book was also made possible by the support of SEP-CONACyT through project 61183, "Coordinate Determination in Ad-hoc and Sensor Networks," within the Basic Science Program.

We are grateful to our friends and former graduate students who assisted us by making their thesis research work available. Among the many who contributed ideas and technical insights through discussions and suggestions are Rafaela Villalpando, Elodia Sanchez, Oziel Hernandez, Raul Torres, Pekka Perala, Michel Z. Antonio, Lluvia G. Suarez, Cynthia Castro, Armando Garcia-Berumen, Enrique Stevens, Eric Baca, Jose I. Bermudez, Martha L. Torres, and Aldo Lopez Gudini.

It was a pleasure to work with the team at Academic Press/Elsevier, and we would like to thank Tim Pitts, Melanie Benson, and Marilyn Rash for their understanding and guidance.

Finally, we thank our families for all of their support during this incredible adventure.

David Munoz
Frantz Bouchereau
Cesar Vargas
Rogerio Enriquez

About the Authors

David Muñoz Rodríguez received a B.S. in 1972, an M.S. in 1976, and a Ph.D. in 1979 in electrical engineering from the Universidad de Guadalajara, México, Cinvestav, México, and the University of Essex in Colchester, England, respectively. He is Senior Member of the IEEE and was formerly Chairman of the Communication Department and Electrical Engineering Department at Cinvestav, IPN. In 1992, Dr. Muñoz joined the Centro de Electrónica y Telecomunicaciones, Instituto Tecnológico y de Estudios Superiores de Monterrey (ITESM), Campus Monterrey, México, where he is the Director. He is also Nortel Telecommunication Chair holder. His research interests include wireless systems and performance analysis.

Frantz Bouchereau received Ph.D. and M.S. degrees in communications and digital signal processing from Northeastern University in Boston in 2004 and 1999, respectively, and a B.S. degree in electrical engineering in 1995 from Instituto Tecnológico y de Estudios Superiores de Monterrey (ITESM), Campus Monterrey, México. He has worked at various research centers in areas such as radiolocalization algorithms, adaptive echo cancellation schemes for ADSL modems, and RF circuit design. His research interests are in the areas of detection and estimation, stochastic and adaptive signal processing, wireless communications, statistical inference in multipath fading channels, and node localization algorithms in wireless networks. From 2004 to 2008 Dr. Bouchereau was an assistant professor in the Electrical Engineering Department at ITESM, and he is currently a senior communications software developer at The MathWorks Inc. in Natick, Massachusetts.

César Vargas Rosales received a Ph.D. in electrical engineering from Louisiana State University in 1996. Thereafter, he joined the Centro de Electrónica Telecommunicaciones, Instituto Tecnológico y de Estudios Superiores de Monterrey (ITESM), Campus Monterrey, México. He has carried out research in the area of personal communication systems about CDMA, smart antennas, adaptive resource sharing, location information processing, and multimedia services. His research interests are personal communications networks, position location, mobility and traffic modeling, intrusion detection, and routing in reconfigurable networks. Dr. Vargas has been a Senior Member of IEEE since 2001.

Rogerio Enriquez-Caldera received a Ph.D. in digital signal processing and communications in 1994 from the University of New Brunswick in Canada. He holds an M.S. degree in theoretical physics and applied mathematics in the area of cosmology. His work includes a GPS-based computer system for location and management of marine and terrestrial transport systems. In the area of wireless ad hoc networks, he has developed a simulator to evaluate package routing in such computer networks. Dr. Enriquez currently works at the Instituto Nacional de Astrofísica, Optica y Electrónica (INAOE), Puebla, México, and is in charge of designing a technology program to promote the use of basic science at innovating companies.

Position Location
Techniques and
Applications

The Position Location Problem

Simply put, the fundamental problem of position location (PL) can be formulated as that of finding or estimating the location, in a two-dimensional (2D) or three-dimensional (3D) space, of a point of interest within a coordinate system constructed using some known references. In general location scenarios, and at the point of interest, a new location is determined bearing in mind the displacement from a previously known reference location. This may imply some direction and inertial estimation. Due to the widespread penetration of wireless systems, we consider devices with transmission or reception capabilities that somehow assist in the location procedure.

In this book, and according to the context of the scenario being discussed and in agreement with terms used in the open literature, we will interchangeably use the terms *point of interest, node, tag, location device, user, subscriber unit,* and *mobile,* among others, to make reference to the location coordinates that need to be estimated. In the same sense, the set of known references used to locate the device will be denoted as references, reference nodes, landmarks, land references, beacons, and anchor nodes.

This chapter addresses the relevance of location information as an important resource that includes other applications for better network planning, development of new location-based services, and fast deployment of assistance services, as well as support of surveillance and security policies. Conflicting criteria in the algorithm design due to PL information requirements and achievable goals (e.g., accuracy, cost, processing load, implementation times) are reviewed. Practical and theoretical limitations, such as measuring errors, impact of a reduced number of landmarks, multihop, uncertain node location, and mobility are introduced. Current trends in sensor and three-dimensional systems are also presented.

1.1 THE NEED FOR PL AND HISTORICAL DEVELOPMENTS

The construction of landmarks can be traced back to the ziggurats on the Shinar plains. These landmarks were meant to be religious as well as geographical references. The usefulness of referencing location to an adopted landmark was soon recognized, and some time in the third century B.C., Eratosthenes of Cyrene devised a system of latitude and longitude.

Today, we accept the Greenwich meridian, the equatorial circle, and mean sea level as references that allow us to specify unique locations on Earth. Global coordinates may be very useful in some applications; however, in other cases, the locations are to be contextualized to other references of regional or local importance. Therefore, location description in a reference system often needs to be translated into another coordinate description suitable for the application at hand. Location information has been an important tool for enabling multiple applications (e.g., community settlement planning; traveling; mining; surveillance; monitoring; military applications; and nowadays the development of location-based services (LBS) and engineering).

In some cases, location information may mean survival. This is the case for a variety of species that have evolved diverse migrating practices and capabilities such as orientation systems based on geomagnetic field sensing. Early humans used star gazing to aid in seasonal migration, farming, and crop cycles. In time, they developed sophisticated astronomic tools, such as the astrolabe, that years later were used by mariners to determine the latitude of a ship at sea by measuring the elevation of a star of known declination in a given season. The presence of the quasistationary North Star in the northern hemisphere allowed the determination of latitude. In contrast, the latitude calculation in the southern hemisphere was more demanding because a fixed visible astronomic reference above the South Pole sky does not exist.

In all cases, longitude had to be calculated from estimated direction and traveled distance, which demanded time references as well. Time measurements also became important references because astronomic observations had to be related to the time of day in a particular season. This led to large navigation almanacs used to cross-check current observations to determine the geographic position of a given site. Although sun dials were among the first time-measurement instruments, they had serious drawbacks such as limited portability and latitude- and season-dependent performance. Thus, water clocks (*clepsydra*) and sand devices soon found a place as timekeeping references. The importance of these devices derives from the fact that they allowed speed measurements since one could now relate travel distance to a given time interval. The magnetic compass became an important navigation tool because it could combine latitude information with travel direction. This combination of techniques prevails as a common practice since no single technique is best for all scenarios, and feasibility must be assessed for each application environment.

Early humans established landmarks to develop relational frameworks that provided location information relative to known references. Their methods relied on major landmarks that were assumed static, a valid assumption for the purpose and time scale of the observations. These developments promoted map construction. Maps used by early sailors were usually distorted and subject to slow iterative processes where current observations were confronted to previous maps and cross-checked with somebody else's map. As location information became critical for commercial, travel, and military purposes, the search for more accurate maps and precise instruments gained great importance and became strategic. This caused the work of navigators to be considered crucial in any journey. Newly acquired and more accurate maps and instruments had such value that they were usually kept secret and/or locked away.

Lighthouses, like those located in Pharos near ancient Alexandria, were among the first human-made landmarks. They were references that could be spotted from afar, allowing proximity estimations of distance. Soon afterward, chains of lighthouses were constructed along common navigation routes for maritime and eventually aviation purposes. In navigation, vessels estimated their location by inferring the direction and traveled distance from the last known landmark. This technique, which became known as *dead-reckoning*, was also applied in early aviation. Limitations of dead-reckoning are due to errors in the parameter estimation process, as well as other undetected disturbances (e.g., tidal and wind drifts). Location updates were applied at the time of the observation of the next known landmark, which was assumed to have an accurate location. At first, without the satellite surveillance technology available today, landmarks' relative locations were quite inaccurate; and, as mentioned earlier, iterative processes were necessary to develop more accurate map tracing and thus more accurate dead-reckoning schemes. Soon after radio systems had been developed, dead-reckoning schemes were also used in applications for the determination of bearings, routes, or locations of vessels and in-flight aircraft.

Modern location systems still rely on basic information such as distance and observable angles, which can be indirectly obtained through multiple related parameters (e.g., field strength, arrival times, arrival time differences, phase differences, and phase variations). Technological advances have contributed not only to a faster acquisition of location-related parameters and improved accuracy but also to the design of new architectures such as cooperative and data-fusion systems.

Advantages, accuracy, and limitations of particular location techniques depend on the costs involved, the number of locations to be determined, technology legacy constraints, autonomy level, processing level, and other considerations that define the operational environment. Good applications are those that achieve an adequate equilibrium between system requirements, technological advantages, and associated costs.

Design of radio-supported localization systems may involve the use of different distance- or direction-dependent measurements. For instance, received signal strength depends, for a given transmitted power, on the distance between a receiver and the radio source. Signal-propagation time is also dependent on distances that can be directly inferred if the propagation speed is known. Angular observations also provide location information about a subject, and in some cases, angular and range measurements are combined to improve performance. Underwater PL schemes are used in submarine positioning applications, exploration, and underwater sensor networks. In these scenarios, PL is usually achieved using acoustic signal transmissions because it is well known that some radio frequencies attenuate very quickly in water.

From the architectural point of view, autonomous systems are meant to determine their own location using natural or human-made references. This was the case with early navigation, where navigators obtained their location relying only on local observations. This is also the case with the global positioning system (GPS), where the receiver is able to infer its own location from the received signals coming from satellites.

Other systems are interactive in the sense that information also flows from the unit to be located to specific processing sites or vice versa. The systems demand synchronization and networking capabilities suitable for centralized or decentralized operation and imply some type of cooperation through information-sharing schemes. Clearly, in military scenarios, alien units will exhibit noncooperative behavior, since they prefer to not be located. In this scenario, information gathered by various beacons is routed to a single central device where it is processed. In other scenarios, radio signals received from friendly units can be processed locally at the reception site in a decentralized manner.

Location information enables multiple applications. In some cases, the applications will depend on achievable accuracy, and in other cases will impose accuracy and architectural requirements. For instance, while military systems are known to be highly accurate, fast at information gathering, and costly, some very simple systems may take whatever information is available and use smart ways to determine location.

The importance of location information and all its applications has contributed to the creation of geography, cartography, and topography disciplines that have evolved over time, incorporating new and advanced technologies. Techniques and applications have gone through a continuous evolution process as society's needs have changed and technology has developed. In the following chapters, we introduce and analyze several position-localization techniques and their sources of error, inherent limitations, and most common applications.

Now that we have briefly introduced the evolution of the need for PL information, the next section discusses some of the main requirements and limitations faced by position-estimation schemes.

1.2 PL REQUIREMENTS AND LIMITATIONS

The ideal goal of PL techniques is to determine the exact position of a tag or node of interest with zero estimation errors. However, it is recognized that this goal cannot be achieved due to cost constraints and, above all, intrinsic limitations.

In the first instance, a location is always related to a reference system that, in the simplest case, may be formed by a set of known landmarks. Note, however, that the relative location of landmarks is usually subject to measuring errors.

Other sources of imprecision are instrument errors. Erroneous measurements in current electronic systems can be attributed to electronic or quantum noise and variation in component parameters, among multiple impairment sources. Measurement errors, noise, and inaccurate land reference positions will have a direct impact on the resolution of PL systems, where resolution can be defined as the accuracy limits that may be achieved with a specific localization scheme. This leads to the consideration of resolution as a fundamental issue.

1.2.1 Resolution

Resolution can be referred to as the ability to discriminate among near locations. Resolution requirements depend on the application. A good example to illustrate localization resolution is the use of postal codes. Usually city names are ambiguous because they can denote settlements in various states and counties. Therefore, the land may be regionalized in disjointed areas denoted by alphanumeric codes. In the United States, for example, a zone improvement plan (ZIP) that consists of five numerical digits was originally developed to improve address resolution. An extended ZIP + 4 code includes the five-digit ZIP code, a hyphen, and then four more digits that allow a piece of mail to be directed to a more precise location. The final address will depend on the street name and house number, which are usually unique to the zone. This sort of partition has been applied for various purposes. For instance, the universal transverse mercator (UTM), as inspired by the military, divides a map into different zones using a grid. Map quantization processes are used in multiple applications (e.g., ad hoc and sensor networks) as will be discussed later in Chapter 4. Resolution will depend not only on the coarseness of a land-quantization scheme but also on other error sources associated with parameter acquisition and measurement procedures.

In radio PL systems, major sources of error are related to signal-propagation phenomena. For instance, reflection, refraction, absorption, and diffraction may cause field-intensity measurements to strongly deviate from ideal nominal values. These variations are usually treated as random and are described using statistical models because their deterministic description is not feasible. As an example, signals at a reflection point exhibit some scattering behavior depending on the roughness of the surface. Here, the reflection coefficient may vary with respect to the material and geometry of the reflecting surface. It is important to note that user mobility renders changes in propagation conditions so that location error is uneven but site dependent.

For a given system configuration, the mean square error is limited by variance of the noise sources. It is common to consider error sources independent of each other, and the rms error is limited by the Cramer-Rao bound as discussed later in Chapter 3 (see also [9, 10]).

Objects in the propagation environment may also cause signal shadowing and reflections that may propitiate a multipath scenario in which reflected copies of a signal travel longer distances than the direct line of sight (LOS). Additive combinations of reflected signals render field intensity variations or fading. These phenomena occur in acoustic as well as in RF scenarios, and their nature and characteristics are highly dependent on the operation frequency band, and even on temperature, for the case of acoustic signals. The obstruction phenomenon prevents signal reception, though the detection of reflected signals may provide some noisy evidence of the radiating object's location. Ground-level variation also produces deviations of expected intensity levels.

In all cases, errors may be reduced through filtering, averaging practices, and multiobservation or redundancy. However, these improvement practices may be costly because they involve deployment of some new reference beacons, such as a base station or a satellite, for redundancy purposes. Further, it is well known that the gain obtained becomes marginal after a certain redundancy level has been reached [9, 10]. This issue will be addressed in more detail in Chapter 4.

Satellite systems offer very high PL resolution; however, few organizations can deploy and operate such a system on their own. Thus, the achievable location information depends on system operator rules and the operating mode. A limiting element of satellite systems is that obstruction of very weak signals prevents any reception at a localization device. Consequently, the availability of the system is limited and/or constrained to regions where GPS receivers may have an LOS to the satellite. The fact that indoor scenarios cannot take advantage of

GPS systems has boosted the search for non-GPS less localization schemes. As will be explained in Chapters 2 and 3, indoor scenarios pose particular propagation problems (e.g., dense multipath environments and shadowing caused by building obstruction) that will directly affect PL resolution. Chapter 7 is devoted to GPS location schemes.

Although high PL resolution is usually desirable, its value may be compromised due to cost limitations and availability. This, for example, may impose limitations on the use of satellite systems in cases where high resolution may not be a requirement. Nevertheless, awareness and ability to analyze the resolution level offered by a specific technological option is of utmost importance for a PL system designer because there will be cases where the achievable PL resolution will not be adequate for a specific application. Position-estimation errors are often referred to in statistical terms and overall system performance can be measured in percentile requirements. For instance, the Federal Communications Commission (FCC) specifies that for at least 95% of emergency calls, a resolution of 300 m must be achievable in a first instance and that this resolution should be increased to 150 m for improved systems. Further, 67% of the calls must be located within an error of 100 m in the initial stage and 50 m in the improved phase.

A majority of location algorithms require measurements of range or relative angles between a node of interest and fixed landmarks. Ranging can be achieved using different signal measurements such as time of arrival (TOA), time difference of arrival (TDOA), and received signal strength (RSS). Relative angles can be inferred from angle-of-arrival (AOA) measurements that may be obtained using antenna arrays to measure the phase difference of the signal as it reaches each array element. TOA-based systems require high clock synchronization between all system components in order for a beacon to know the exact time when a signal was transmitted by a source. It may be possible to synchronize all the beacons on the network, but it is quite difficult, due to hardware costs and system limitations, to synchronize the node of interest. Poor synchronization is one of the main sources of error in TOA-based systems along with multipath, since several unresolvable paths arriving closely in time will impede the differentiation of the first TOA from the rest. Of course, low signal-to-noise ratio (SNR) values also limit the resolution of TOA measurements.

Typical TOA-based range estimation methods may involve the transmission of a wideband pseudo noise (PN) sequence and subsequent correlation of the received signal with this sequence. Shadowing could also be a cause for attenuation of the LOS signal. When this happens, a contiguous, subsequent strongest non-LOS signal arrival may be chosen as the first arrival and this translates into range estimation errors. High-resolution TOA measurements require signals with large bandwidth in order to be able to resolve the multipath at the reference node end. It is known that resolution of TOA measurements is inversely proportional to the signal's bandwidth (BW).

Several signal-processing techniques can be applied to achieve better resolution without increasing signal BW. For instance, maximum likelihood (ML) delay estimation can provide greater delay resolution. ML schemes require either estimation of amplitude and phase parameters of the multipath channel or averaging of a likelihood functional over a joint distribution. Improved performance can also be achieved by using methods based on noise and signal subspace decomposition. These methods have been shown to achieve time resolutions of a fraction of the sampling interval without increasing signal BW [13]. In certain scenarios, multipath arrivals may be highly correlated. In this case, the subspace methods fail to estimate path arrival times accurately unless a path decorrelation technique is applied. Estimation of the PL of a source located close to a beacon becomes difficult to achieve with TOA measurements due to the fact that the time of travel may be smaller than the time resolution available at the observation sites.

TDOA systems avoid the requirement of clock synchronization at the point of interest or tag-end by considering the different arrival times of signals that originate at two distinct reference points. The time difference calculation effectively cancels out time synchronization errors at the tag. Notice that TDOA-based positioning schemes still require clock synchronization between all the beacons in the system. Measurements of TDOA also involve the correlation of transmitted and received signals to determine the arrival time of the strongest correlation peak. Thus, excluding synchronization issues, the sources of error of TDOA-based systems are basically the same as the ones described for TOA schemes.

AOA-based systems usually require an array of arrival sensors in order to detect phase differences arriving at each array element and to infer arrival signal direction. Resolution of AOA measurements is limited by the SNR, the number of sensors in the array, and the separation between these sensors; the last two characteristics define what is known as *array aperture*. Similar to the TOA-estimation scheme, array-processing techniques based on noise and signal subspace decomposition allow an increase in resolution without the need to increase the number of elements in the array. It has been shown that certain subspace methods are able to increase angular resolution to values smaller than the beamwidth of the array without the need to increase the array aperture.

AOA-estimation schemes are usually thought of as costly, computationally expensive, and bulky because they may require large antenna arrays. Note, however, that technology development trends in wireless networks favor the use of higher-frequency bands that will allow the implementation of smaller antenna arrays on the network nodes. Moreover, recent developments propose computationally simple AOA-estimation schemes that require small antenna arrays with a reduced number of elements (see, for example, [2–5]). Multipath is one of the main causes of errors for AOA measurements because it becomes difficult to resolve the direction of arrival of multiple signals with closely spaced angles at the sensor array.

Some advantages of AOA-based localization systems are that they do not require synchronization at the beacons or at the tags, and the number of beacons required for the location process is lower than what is required by TOA or TDOA schemes. A disadvantage is that accuracy degrades as the distance between the tag and the antenna sensors increases. Another source of ambiguity occurs when a tag to be located lies on the straight line connecting two beacons. This situation can be overcome by using additional beacons.

RSS-based systems require the simplest hardware (only a power detector), which is readily available in several applications such as WiFi, Zigbee, and Bluetooth chipsets. These schemes do not require synchronization.

Location systems based on RSS are highly dependent on the propagation model used to infer distance. Further, it can be shown that the estimation performance of RSS-based distance inference decreases as the distance between a source and the receiver increases. RSS systems perform very well in short-range scenarios where TOA systems are limited due to finite time resolution.

RSS systems are very sensitive to shadowing and to non-LOS scenarios since, in these cases, signal power decreases considerably causing large estimation errors. These systems are also highly affected by local variations on the average received power (i.e., to small-scale fading). Note however that these systems are more robust to multipath because they do not rely on timing information [12]. RSS-based schemes are also sensitive to low SNRs and variations in attenuation due to weather conditions (e.g., rain, fog, or snow).

Proximity, as seen by a node of interest, refers to the knowledge of the position of its neighboring nodes. Proximity-based PL-estimation systems are usually called range-free because no ranging or AOA-estimation measurement is required in the process. A simple example of a proximity-based system is the centroid method, where a node of interest estimates its position as the centroid of a polygon formed by the concatenation of neighboring nodes' coordinates. Proximity-only methods usually achieve poor resolution performance; however, these schemes are not expensive and are simple to implement. Therefore, proximity information is typically used together with other ranging metrics as a tool to improve PL accuracy.

Proximity is very useful in scenarios where, more than knowing precise location information, one needs to understand the functional relationships between devices in a network. For instance, knowledge of the closest neighboring nodes may be useful in designing efficient routing algorithms in reconfigurable networks. Also, in cases where direct connectivity to fixed land references or beacons is not available, proximity information collected throughout the network may be transmitted by neighboring nodes to a node with unknown location so that it can estimate its position or at least obtain a rough picture of current network topology.

Hybrid PL schemes combine the use of two or more parameter measurements to improve performance. For instance, in the case of a mobile near the base station, changes in field strength due to location variation may be significant while TOA resolution is limited for short distances; the use of hybrid TOA and RSS measurements increases range-estimation robustness.

1.2.2 Fundamental Scenarios for PL

Whenever position-estimation needs to be carried out, we can construct a generalized scenario that could explain the fundamental tasks that need to be performed independently of the type of technology and application. All these scenarios involve some sort of communication between one or more devices that need to be localized and a set of reference devices with an assumed location. Usually, a tag is a device with unknown location that is capable of transmitting or receiving a signal to or from nodes that lie within its coverage area (Figure 1.1). The types of signals vary depending on the specific application, but they are usually radio waves, optical, ultra-wideband (UWB), or acoustic signals. Some tags may have the capability to communicate with other nodes and to measure or infer certain parameters such as RSS, TOA, TDOA, AOA, and proximity. Others may act as simple backscatterers that reflect whatever signal they receive. Some active backscatterers may even change the signal frequency, phase, or randomize retransmission instant [1] to avoid collisions with other signals.

A beacon is a device usually capable of transmitting or receiving a signal from other beacons, nodes, and tags. Beacons are usually, but not necessarily, more computationally powerful and their most important feature is their known location, at least within a given confidence interval. Beacons may also have the capability to measure or infer certain parameters such as RSS, TOA, TDOA, AOA, and proximity. Beacons may be fixed devices placed by a user at specific coordinates, such as location mobile units (LMUs) in a GSM system that can be located at the base-station site or in alternative places. Beacons can also be mobile nodes that are aware of their location thanks to a positioning device such as a GPS receiver.

Finally, both beacons and tags have a defined finite reachability radius that limits their communications capabilities to given coverage zones. Hence, there may be cases when several beacons exist in a network, but a tag may not communicate with any of them due to its limited coverage range. Nodes within the coverage zone of a tag or beacon are usually referred to as neighboring nodes.

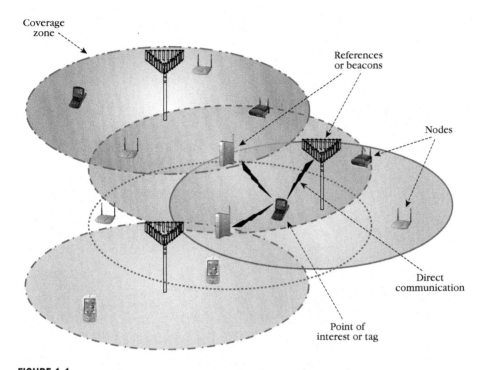

FIGURE 1.1

General scenario for the PL single-hop case.

The ultimate goal of a position-estimation algorithm is to combine measurements of range or relative angles between one or more tags and a number of beacons in order to infer, with as much accuracy as possible, the position of these tags. These ranges and relative angles are parameters that may be extracted from RSS, TOA, TDOA, AOA, and proximity measurements.

The simplest scenario, referred to as single hop, consists of a single tag trying to estimate its position by communicating directly with multiple beacons whose coordinates are fixed and exactly known. In this description, the word "directly" implies that beacons are within the tag's coverage zone or vice versa (see Figure 1.1). An example of this type of scenario is the cellular position location system where a mobile may be localized from range and AOA measurements obtained at different base stations in its proximity. Another example is the GPS system where a receiver estimates its position by performing ranging measurements from at least four satellites.

A variation of the preceding scenario, referred to as multihop, may arise when the tag's coverage zone is not large enough to reach a beacon; then the tag may establish an indirect communication with one or more beacons via multiple

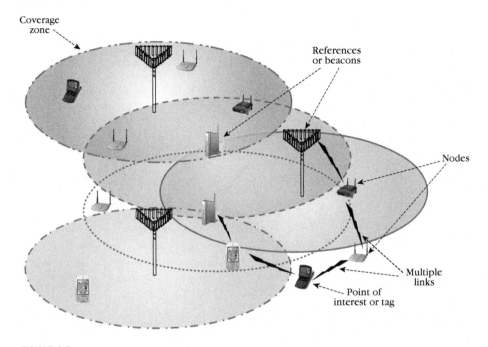

FIGURE 1.2

General scenario for the PL multihop case.

communication links (multiple hops) established with neighboring nodes, as shown in Figure 1.2. These types of scenarios arise in reconfigurable mobile networks as well, such as ad hoc and sensor networks.

Geometric or statistical multilateration techniques are commonly used to combine range, AOA, and proximity measurements to obtain position estimates of a tag in the vicinity of three or more beacons. Accuracy of the position estimates will be affected by the number of available beacons, node density, the type of routing algorithms, and the coverage area of the network nodes, among others.

Cooperative algorithms are a popular type of multihop position-estimation scheme. In these algorithms, data, such as beacon coordinates or measurements of proximity, range, and AOA between network nodes, are propagated iteratively throughout one or more sectors of a reconfigurable network so that, after a number of iterations, a tag that is initially away from the neighborhood of any beacon may receive this information and use it to estimate its position. During the iteration process, intermediate nodes may also be able to estimate their positions and may act as newly acquired beacons if desired. Depending on the size of the network sector and the number of intermediate nodes involved in the data propagation process, cooperative algorithms may be classified as localized or nonlocalized.

In a typical nonlocalized cooperative algorithm, a node that is in the vicinity of more than one beacon may be able to estimate its position using multilateration techniques based on (possibly combined) measurements of proximity, range, and AOA. When this node accomplishes its position-estimation procedure, it may act as a new beacon (with reduced position accuracy) and assist its neighboring nodes in their own position-estimation processes.

Newly generated beacons are iteratively propagated through the network until all nodes locate themselves or until sufficient information reaches a tag so that it can be localized. The number of necessary intermediate nodes that need to be involved in the cooperative process to reach a tag will depend on the network density, the nodes' reachability radii, and the position and density of initially available beacons. Depending on these parameters, and on the position-estimation algorithm itself, the region occupied by the intermediate nodes may expand over a large sector or even the complete network area. Nonlocalized methods are highly sensitive to beacon densities and may fail to converge in places where these densities are sparse. Further, error propagation of newly created beacon nodes may become an issue in large network scenarios.

Localized cooperative methods, on the other hand, do not require the tag to be in one-hop proximity of anchor nodes; the only requirement is that it has connectivity, via multihop routes, to three or more beacons. Thus, the only intermediate nodes involved in the position-estimation process are the nodes that form those routes.

Cooperative positioning algorithms may also be classified according to the computational load at each node. Noncentralized (or distributed) algorithms distribute the computational load evenly across the network's nodes, taking advantage of all the computational power available at the network. On the other hand, data collected by all the nodes in the network may be transmitted and stored at a single computationally powerful node capable of solving complex optimization problems in order to infer position of a node of interest, a group of nodes, or all the nodes in the network.

Finally, PL scenarios may be categorized, according to the infrastructure available to obtain position information, as exogenous or endogenous. In exogenous systems, infrastructure deployment is specifically oriented toward a PL application. This is the case of GPS and long-range navigation (Loran) systems where a location-oriented infrastructure has been clearly deployed.

On the other hand, endogenous systems are those where position-estimation is based on infrastructure that has not been deployed specifically for localization purposes. Here, PL algorithm designers use the various systems' capabilities in ingenious ways to obtain position information. This is the case with cellular systems that base mobile position-estimation on received field strength, propagation delay, and observed angles of arrival, using only the cellular system's infrastructure and the system's networking facilities, which obviously have not been designed specifically for localization purposes. Another example is Internet protocol (IP)-based localization, where a user may be identified through the declared IP address of the nearest server [6, 7].

1.3 TERRESTRIAL AND SATELLITE SCENARIOS

In the search for PL solutions, ingenious ways to estimate location have been proposed. Several methods suffer from accuracy degradation as distance increases between a point of interest and a reference. Hence, a good question is how to attack the PL problem when considering large distances. Further, consider location of a node in the middle of the sea. What would the terrestrial "sea" references be? Here, a new type of reference is needed: Celestial objects are the obvious solution.

In all possible terrestrial solutions for PL one can imagine two possibilities: One is positioning with respect to a generally accepted reference frame; the other, positioning with respect to another arbitrary but known point or a set of points. In the first case, a coordinate system (commonly defined in terms of what is known as the datum) is a must. Several people can agree on one or more datums, but all have to be well defined with respect to the Earth. This positioning can be referred to as point or absolute positioning. In the second case, one arbitrary point may be taken as the origin of a local reference frame. This positioning is called relative positioning or differential positioning.

In either case, positioning may involve static or kinematic scenarios. When kinematic scenarios arise, they are commonly referred to as *navigation. Datum* is a fundamental concept; a very popular and therefore classical datum nowadays is the geoid, the surface that defines the actual shape of the Earth and allows the definition of latitudes and longitudes. Height, on the other hand, is measured with respect to sea level, which is best fitted by the isosurface of the terrestrial gravity field. The geoid is usually substituted with an ellipse that is used to generate an ellipsoid as a reference for latitude and longitude coordinates.

One should realize that when using celestial objects as references, they are in different reference frames or coordinate systems than the ones described above, and that transformations must be considered. Highly accurate PL using celestial objects involves complicated coordinate transformations and, although computers can simplify such tasks, inherent larger uncertainties decrease the region of confidence for any PL scheme. In this sense, satellites are a link between terrestrial and celestial reference systems in which uncertainties are reduced due to the controlled positioning of the satellites in the Earth's orbit.

Satellite PL was developed in the latter half of the 20th century. It started with early radio ranging and direction finding, and then evolved into a more proper PL scheme with the appearance of the transient system and, in recent times, into the well-known global positioning system. Various techniques are used in satellite PL and are discussed throughout this book. The main disadvantage of satellite systems is the lack of 100% land coverage. On the other hand, the appearance of wireless computer networks as well as mobile telephony networks allows nearly 100% coverage in continental land masses.

In cellular systems, subscriber locations can be handled at various levels. In the first step, information, where the subscriber is on a region, will be used to page the mobile units. In this case, records are maintained and updated indicating the set of base stations that may be able to reach the mobile. A set of base stations forms a location area where their associated transmitters emit a location area identifier (LA-ID). When the subscriber unit receives a new LA-ID, it sends a notification via its current base station in order to update the LA database. In a further step, when a connection is established, information about the base station (or sector) used to reach a mobile is used to reduce location options to that base station's coverage area. Thus, while hot spots may permit a quite accurate location estimation, the location uncertainty will be large in macrocellular scenarios.

To improve location accuracy of a mobile connected to a base station in a given cell, measurements may be performed to determine its range and direction to that base station. Although these measurements will be noisy (as has been previously explained), redundancy will tend to reduce noise effects. A common way to achieve redundancy is measuring distance and/or angles of arrival from multiple base stations. These multiple measurements may then be combined using a geometric or statistical multilateration approach. The most common geometric approach involves finding intersections of circles with centers that correspond to base station positions and with radii that correspond to the estimated ranges. In the absence of noise, ambiguities are prevented with the use of at least three circles. The circle intersection approach may be extended to alternate algebraic representations with the use of other conical sections. For instance, the locus of points, whose distance difference to two base stations is constant, defines a hyperbole, and, if three base stations are used, one may obtain several hyperbolas with an intersection that is meant to denote the subscriber location.

This technique is useful in cases when direct range finding is not possible, but the range difference with respect to two base stations is available. This is the case of TDOA where there is no knowledge of the transmission instant but reception times at two synchronized base stations is used to obtain the distance difference. The presence of noise brings back the ambiguity problem, and the problem can be tackled by heuristics, maximum likelihood estimation, or other (possibly suboptimal) statistical estimation techniques. A wide variety of geometric and statistical multilateration methods, as well as heuristics approaches, will be discussed throughout this book.

Note that scenario descriptions have typically been related to planar location in a 2D space. In the case of 3D scenarios, additional observation references are required, and typical position-estimation–related measurements, such as AOA, must be extended to consider elevation.

1.3.1 Mobility

Mobility knowledge, or tracking, is important in several systems in order to assist resource allocation planning or to predict future states of a population or individual. Mobility measurements, such as velocity, are usually addressed in terms of Doppler shift [11], which concerns signal frequency changes seen at the observation point. Precise determination of the velocity vector demands several observations and also knowledge of the radio source location. From our perspective, mobility may be defined as changes in location in a given period of time. In this sense, a mobility indicator may be obtained by comparing location positions at consecutive moments; large position changes will correspond to high mobility. Observation of position changes over time can be conducted on an individual basis or for a population as a whole. In this last scenario, the processing load will become a constraint as position-estimation of several tags must be completed at each time interval. In the case of centralized systems that contain a large number of mobiles, processing overload may be a limiting factor for high-mobility scenarios. For this reason, many systems avoid dealing with the location of all the individuals in the network. Instead, these systems perform localization and mobility inferences by request. For instance, in the E-911 scheme, subscribers in distress may trigger a location procedure and tracking is done on an individual basis rather than for the whole population.

In the case of a large number of individuals, decentralized systems provide a better alternative, as the processing overload phenomenon will not occur. For instance, in GPS, each individual unit is able to obtain location coordinates on its own. Nevertheless, how the location information obtained by a single GPS receiver is fed to a central office, for more sophisticated applications, needs to be addressed.

In many situations, mobility occurs along streets and other car paths. This allows location to be confined to a reduced number of geographic options. Although simplified scenarios may not be applicable to pedestrians, the use of subspace search can be used with some simplification. For instance, if a coverage area is divided into a discrete number of regions or spots, the location problem may reduce to determining the spot where a mobile stands. This procedure introduces a quantization error that depends on the coarseness of the discretization process.

Throughout this book, mobility issues within the PL estimation problem, such as tracking, accuracy effects, required network architectures, and modeling will be introduced and discussed.

1.4 CURRENT AND POTENTIAL APPLICATIONS

Location prevails as a piece of crucial information for decision-making processes. For instance, in the European Union, of approximately 40 million emergency calls originated from mobile phones, emergency services could not be dispatched in response to 2.5 million requests because of insufficient location information. PL information about mobiles generating emergency calls is of utmost importance since, in many cases, timely response increases survival chances and reduces catastrophic escalation, as in firefighting operations.

In POTS (plain old telephone service) systems, call locations are directly, and simply, associated to an address known to the network operator. However, changes in communication practices and new network architectures pose new technological challenges when trying to provide service according to subscriber expectations. In some countries, approximately 50% of the calls are generated from mobile units whose whereabouts are meant to be unrestricted. Yet, assistance services must be provided when requested. Under emergency situations, callers may be emotionally distressed or unfamiliar with the environment and thus not be able to provide correct directions to their location to the emergency call center. Thus, automatic location information (ALI) systems become a key service that allows improving accuracy and time response to emergency calls. This has motivated the FCC in the United States and ETSI in the European Union to set PL policies in order to promote technological developments that can cope with location requests in current telecommunication scenarios.

Among other applications, we can mention road monitoring and traceability. These applications are controversial because they involve conflicting interests of privacy and security, and will certainly require regulations to promote and limit their use for commercial and advertising practices as well as for crime prevention and surveillance. Automatic crash notification (ACN) is among the systems under development in which location is a core information element. For instance, location information can be used to determine mobility practices of subscribers on an individual basis or as the aggregate behavior of a population. Mobility patterns will allow the development of models applicable in road traffic planning, telecommunication resource management, and new commercial and marketing practices.

Other applications relate to point-of-interest services, where the subscriber's affinity information is provided to a customer depending on his/her current location. Information may be provided in either "push" or "pull" mode. For instance, a person may be using a location-aware search engine on the Internet where search results will be highly correlated to his or her current geographic location. Navigation systems have been in place enabling the delivery of tailored information (e.g., recommended route, estimated traffic speed, etc.) according to current location, prevailing traffic conditions, and travel destination of a user. In the area of public and freight transportation, route surveillance is advantageous for assisting in fleet management. For instance, a taxi driver can be advised on the large number of empty cabs in a particular area, and the need of units in a nearby region. A similar principle may be applicable to buses, ambulance, fire brigades, and patrol squads.

While reliability and accuracy are prime requirements for positioning schemes, availability and system autonomy also promote the existence of alternative or redundant systems. This policy has ensured the near future of enhanced Loran systems, the development of alternative satellite systems, and the operation of hybrid systems that combine existing technologies, as in the case of locating television systems (LTS) where TV signals will be used to provide location information [8].

Ad hoc and reconfigurable networks have been used extensively in military environments that mix different technologies. Ad hoc networking applications in other areas are being developed jointly with sensor networks. From the perspective of location, ad hoc networks present additional challenges as there are no direct links to fixed references. The fact that end-to-end connection is established after several consecutive hops contributes to location uncertainty. Precision requirements are application dependent, and in some cases proximity to other nodes whose location may also be uncertain may be adequate to generate a broad image of the different nodes in a network.

The ever-growing popularity of sensor and wireless networks is certainly potentiating a vast number of applications. Diverse types of wireless network infrastructures are being deployed throughout entire residential areas, city commercial centers, university and company campuses, hospitals, amusement parks, and restaurants. Wireless networking devices constitute the main infrastructure to be used for wireless location algorithms. In addition to emergency services, several other applications can be envisioned with wireless-network-based position-estimation schemes such as the following:

- Real-time display of self-location information. For example, a user walking across a university or company campus holding a pocket PC that shows the campus map and the real-time position of the user in this map

- Real-time inventory systems where expensive assets can be counted automatically and localized

- Monitoring of position-sensitive environmental variables

- Monitoring and mapping of potentially hazardous zones in disaster areas by deployment of unmanned sensing vehicles prior to the entrance of rescue teams

- Intelligent warehouse systems to optimize stock product flows and logistics

- Patient locators in mental health–care institutions, newborn locators in maternity rooms, or child locators in amusement parks or shopping malls

- Real-time shopping mall guides where users with a personal digital assistant (PDA) or pocket PC can read about sales and featured products of the stores they are approaching while walking through the mall

- Real-time amusement park, museum, or tourist guides that display information about attractions, artwork, or city landmarks according to user location

- Mobile yellow pages

- Mobile databases where information retrieval and file accessibility depend on the geographic location of a user

- Location-aware gaming systems

- Location-sensitive billing in cellular systems

- Location-aware marketing systems that can track location and density of potential clients within a specified area

- Location-aware wireless access security systems where files, information, or system access may or may not be available to a user depending on current geographic position

- In the area of communication systems, applications that include location-aware routing in reconfigurable networks, spatial diversity techniques, knowledge-based systems, distributed learning, cognitive radio, and user density and mobility detection with applications to real-time network reconfiguration and planning

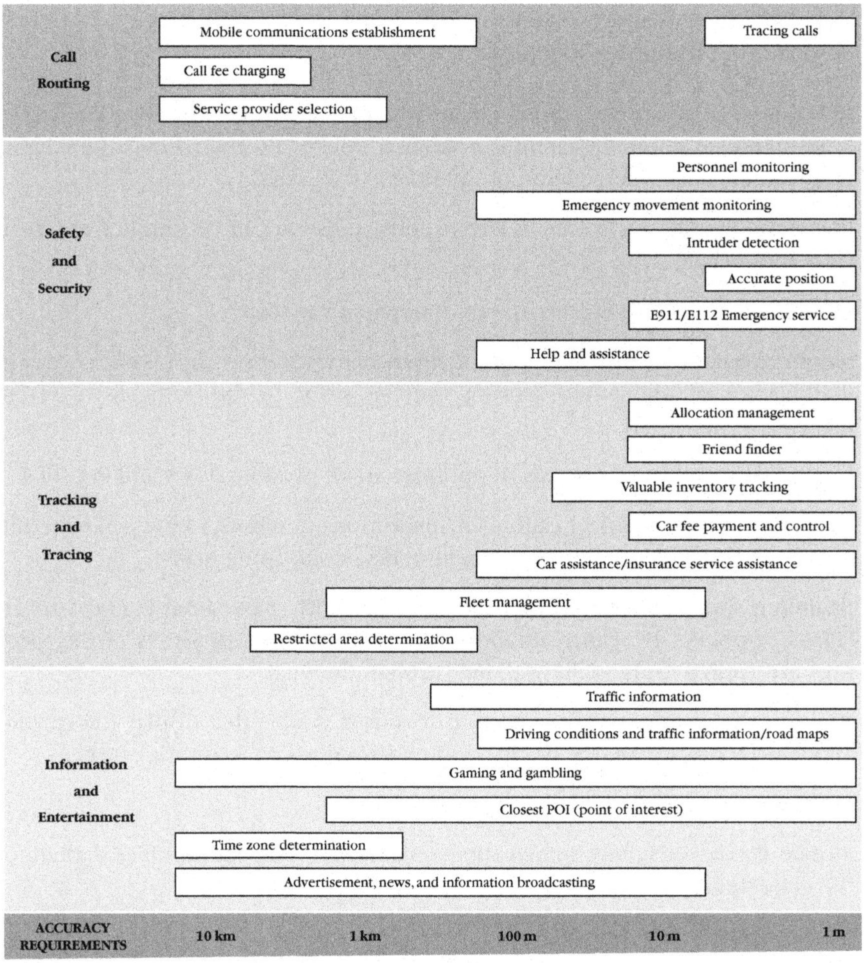

Call Routing	Mobile communications establishment				Tracing calls
	Call fee charging				
	Service provider selection				
Safety and Security				Personnel monitoring	
			Emergency movement monitoring		
				Intruder detection	
				Accurate position	
				E911/E112 Emergency service	
		Help and assistance			
Tracking and Tracing				Allocation management	
				Friend finder	
			Valuable inventory tracking		
				Car fee payment and control	
			Car assistance/insurance service assistance		
		Fleet management			
	Restricted area determination				
Information and Entertainment			Traffic information		
			Driving conditions and traffic information/road maps		
	Gaming and gambling				
			Closest POI (point of interest)		
	Time zone determination				
		Advertisement, news, and information broadcasting			
ACCURACY REQUIREMENTS	10 km	1 km	100 m	10 m	1 m

FIGURE 1.3

Relationship of precision and service classes.

Figure 1.3 presents various service classes and their precision requirements. This list is by no means exhaustive as the application of location is driven by ingenuity in a changing technological environment characterized by decreasing technology cost. Figure 1.4, presents a graphical representation of the compromise that exists between accuracy and technology with respect to PL precision.

Location-based services demand, in addition to location information, user identification procedures that validate the information flow through authentication of user and/or mobile equipment.

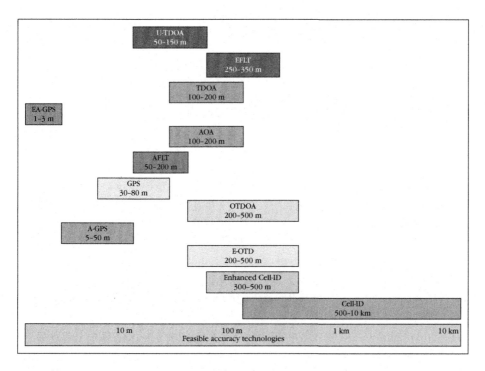

FIGURE 1.4

Technology and location precision.

In the following chapters, we introduce the analysis of the PL problem in a formal manner in order to provide the necessary analytical tools to formulate the algorithms and methodologies used to counteract environmental, infrastructure, and cost limitations.

REFERENCES

[1] J. Werb, C. Lanzl, Designing a positioning system for finding things and people indoors, IEEE Spectrum Magazine, September (1998) 71–78.

[2] H.G. Schantz, Smart antennas for spatial RAKE UWB systems, IEEE International Symposium on Antennas and Propagation, 3 (2004) 2524–2527.

[3] J. Yli-Hietanen, K. Kalliojarvi, J. Astola, Low-complexity angle of arrival estimation of wideband signals using small arrays, in: Proceedings of the 8th IEEE Signal Processing Workshop on Statistical Signal and Array Processing, June (1996) 109–112.

[4] Z. Popovic, C. Walsh, P. Matyas, C. Dietlein, D.Z. Anderson, High-resolution small-aperture angle of arrival detection using nonlinear analog processing, IEEE International Microwave Symposium Digest, 3 (2004) 1749–1752.

[5] W. Menzel, A. Gronau, V. Maiyappan, W. Mayer, Small-aperture, high-resolution beam-scanning antenna array using nonlinear signal processing, European Conference on Wireless Technology, October (2005) 435–438.

[6] M. Mosmondor, L. Skorin-Kapov, M. Kovacic, Bringing location based services to IP multimedia systems, IEEE MELECOM (2006) 746–749.

[7] M. Mosmondor, L. Skorin-Kapov, R. Filjar, Location conveyance in the IP in next generation teletraffic and wired/wireless advanced networking, Lecture Notes, 4003 (2006) 96–107.

[8] M. Rabinowitz, J.J. Spilker, A new positioning system using television synchronization signals, IEEE Transactions on Broadcasting, 51 (2005) 51–61.

[9] H.L. Van Trees, Detection, Estimation, and Modulation Theory, John Wiley & Sons, 1968.

[10] J.M. Wozencraft, I.M. Jacobs, Principles of Communication Engineering, Waveland Press, Inc., 1990.

[11] T.S. Rappaport, Wireless Communications: Principles and Practice, 2nd edition, Prentice Hall, 2002.

[12] F. Bouchereau, D. Brady, Bounds on range-resolution degradation using RSSI measurements, in: Proceedings IEEE International Conference on Communications, France, June (2004) 3246–3250.

[13] F. Bouchereau, D. Brady, C. Lanzl, Multipath delay estimation using a superresolution PN-correlation method, IEEE Transactions on Signal Processing, 49 (5) (2001) 938–949.

Signal Parameter Estimation for the Localization Problem

2

The vast majority of existing positioning systems involve the use of different distance or direction-dependent measurements. For instance, received signal strength (RSS) is dependent, for a given transmitted power, on the distance between a receiver and the transmitting source. Signal propagation time, usually referred to as time of arrival (TOA), is also dependent on distance, which may be directly inferred if the medium propagation speed is known. Angle of arrival (AOA) observations also provide location information about an emitter, and in some cases, angular and range measurements are combined to improve performance. Clearly, the accuracy of a position location scheme will strongly rely on the accuracy of the corresponding range or angular measurements, and for this reason, correct understanding of their measurement and estimation techniques and of their limitations becomes of utmost importance.

This chapter is devoted to presenting some popular AOA, TOA, and RSS measurement/estimation techniques, and will provide the reader with an overview of the different phenomena that limit the accuracy of these measurements. Theoretical accuracy bounds will also be presented, for some specific scenarios, in order to provide more insight on how the different transmitted signal, channel, and receiver parameters affect range and direction measurements.

2.1 AOA MEASUREMENTS

The angle at which a signal arrives at a sensor, such as a hydrophone or an antenna, yields important information about the location of the signal source. As will be explained in Chapter 3, two error-free AOA measurements obtained at two different sensors located at different points in space will be enough to estimate the exact position of the signal-generating source in a 2D scenario. In this section, we discuss several techniques to measure the AOA of a signal, or a group of signals, in the presence of several impairments such as sensor noise, unknown sensor response, limited sensor observations, and signal propagation distortion such as attenuation and multipath.

Most of the existing AOA estimation techniques rely on observations of signals at the output of multiple sensors spaced by fractions or a few wavelengths, and deployed with a specific geometrical arrangement. These sensor schemes are usually referred to as sensor arrays. A signal $s(t)$ impinging on the array with an angle different than 90 degrees (with respect to the array axis) will reach individual sensors at slightly different times; thus, sensor outputs will contain time- and phase-delayed versions of the original signal. Array processing techniques take advantage of the time and phase delay that exists at the sensor outputs and attempt to extract the corresponding embedded angle of arrival information.

In this chapter we will focus on array processing techniques for AOA estimation. We will present the AOA estimation algorithms and then discuss their strengths and weaknesses and, specifically, their sensitivity to noise and channel impairments. We will also present theoretical bounds on performance for the case of arrays with additive white Gaussian noise (AWGN).

2.1.1 The Uniform Linear Array Model

Consider the problem of solving the directions of arrival of multiple signals produced by narrow band radiating uncorrelated sources in the presence of background noise. We assume D signals incident on a uniform linear array of M sensors from angles $\{\theta_i, i = 1, 2, \ldots, D\}, \theta_i \in [-90°, 90°]$. It is also assumed that the array is in the far field of any of the sources such that the impinging signals are planar. Let us first concentrate on the signal generated by the i-th source, as shown in Figure 2.1, and its corresponding output at the m-th sensor. Note that a wave arriving at sensor m travels an extra distance of $d \sin \theta_i$ meters when compared to the distance traveled to reach sensor $m - 1$. This means that signals at contiguous sensors have a time delay of $d \sin \theta_i / u$ seconds, or a phase delay of $k_i = \omega_c d \sin \theta_i / u$, where u is the propagation speed in meters per second, and ω_c is the carrier frequency of the narrow band signal. To avoid ambiguities in the spatial sampling of the array, the phase delay should be limited to $|k_i| < \pi, i = 1, 2, \ldots, D$. This condition is achieved for any θ_i, $i = 1, 2, \ldots, D$, if the separation between the array sensors is limited to $d < \frac{\lambda_c}{2}$, where $\lambda_c = 2\pi u / \omega_c$ is the wavelength of the impinging narrow band signals.

If we reference the sensor output phases with respect to the first sensor (i.e., sensor one is considered to have zero phase delay), then the output at the m-th sensor due to the i-th source may be written as

$$r_m(t) = H_m(\omega_c, \theta_i)e^{-j(m-1)k_i}s_i(t) + \eta_m(t), \tag{2.1}$$

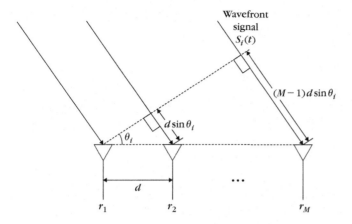

FIGURE 2.1

Uniform linear antenna array.

where $H_m(\omega_c, \theta_i)$ is the frequency response of the m-th sensor evaluated at the center carrier frequency. Note that it has been emphasized that this sensor response is a function of the signal's angle of arrival. This response is mainly determined by the inherent sensor characteristics such as its gain, directivity, sensitivity pattern, and so on. The term $\eta_m(t)$ is additive stationary white Gaussian noise at the m-th sensor with power equal to σ^2, which may correspond to thermal noise generated by the sensor's hardware. This noise is assumed to be uncorrelated with the signals.

Let us now consider all D signals impinging on the array to obtain the field received at the m-th sensor at time t as

$$r_m(t) = \sum_{i=1}^{D} H_m(\omega_c, \theta_i)e^{-j(m-1)k_i}s_i(t) + \eta_m(t); \quad m = 1, \ldots, M, \quad (2.2)$$

For any real number θ, we define the corresponding *steering vector* as

$$\mathbf{a}(\theta) = \left[H_1(\omega_c, \theta)\, H_2(\omega_c, \theta)e^{-jk}\, H_3(\omega_c, \theta)e^{-2jk} \ldots H_M(\omega_c, \theta)e^{-(M-1)jk} \right]^T,$$
$$(2.3)$$

$$k = \frac{\omega_c d}{u}\sin\theta.$$

Clearly this vector depends on θ through the phase delay k, and the sensor responses $H_m(\omega_c, \theta)$. The signals received by the M sensors can then be written as

$$
\begin{bmatrix} r_1(t) \\ r_2(t) \\ \vdots \\ r_M(t) \end{bmatrix} = [\mathbf{a}(\theta_1)\,\mathbf{a}(\theta_2)\cdots\mathbf{a}(\theta_D)]
\begin{bmatrix} s_1(t) \\ s_2(t) \\ \vdots \\ s_D(t) \end{bmatrix} +
\begin{bmatrix} \eta_1(t) \\ \eta_2(t) \\ \vdots \\ \eta_M(t) \end{bmatrix}, \quad (2.4)
$$

or more compactly,

$$\mathbf{r}(t) = \mathbf{A}\mathbf{s}(t) + \boldsymbol{\eta}(t). \tag{2.5}$$

Matrix $\mathbf{A} = [\mathbf{a}(\theta_1)\,\mathbf{a}(\theta_2)\,\cdots\,\mathbf{a}(\theta_D)]$ is usually called the *array manifold*, and its columns are referred to as *steering vectors* since, as will be shown, they may be used to "steer" the array response in any desired direction. It is assumed that for any set of distinct parameters $\theta_i, i = 1, 2, \ldots, D, M > D$, the columns of $\mathbf{A}$ are linearly independent. The array manifold may be expressed in a closed analytical form or measured through field calibration procedures. Actually, a major source of error in AOA estimation methods consists precisely of the incorrect calibration of the array sensors, which basically translates into the incorrect modeling of the sensor's transfer functions [44].

In many scenarios, it is possible to assume that all M sensors are omnidirectional and identical; then $H_m(\omega_c, \theta) = constant\ \forall\, m, \theta$. In this case, without loss of generality, the constant gain may be embedded in the amplitudes of signals $s_i(t)$ to obtain an array response as in Equation (2.4), but with simplified steering vectors

$$\mathbf{a}(\theta) = \left[1\, e^{-jk}\, e^{-2jk}\, \ldots\, e^{-(M-1)jk}\right]^T. \tag{2.6}$$

Returning to the general case, from Equation (2.5) the spatial array correlation matrix can be written as

$$\mathbf{R} = E\{\mathbf{r}(t)\mathbf{r}(t)^H\} = \mathbf{A}\mathbf{S}\mathbf{A}^H + \mathbf{R_n}, \tag{2.7}$$

where $\mathbf{S}$ and $\mathbf{R_n}$ are the correlation matrices of $\mathbf{s}(t)$, and $\boldsymbol{\eta}(t)$, respectively. The noise is assumed spatially and temporally uncorrelated so that $\mathbf{R_n} = \sigma^2\mathbf{I}$, and the signal correlation matrix is assumed nonsingular and not necessarily diagonal, implying that signals may be partially correlated. Note that if the signals are fully correlated then they are said to be coherent and indeed, in this scenario, $\mathbf{S}$ will become singular.

In many situations it will be convenient to weight the sensor outputs and add the resulting products to obtain an overall array response $r_B(t)$. This is achieved with a so-called beamformer whose structure is shown in Figure 2.2. The resultant beamformer output signal $r_B(t)$ may be written as

$$r_B(t) = \mathbf{w}^H\mathbf{r}(t), \tag{2.8}$$

where $\mathbf{w} = [w_1 \ldots w_M]^T$ corresponds to the vector of weights of the beamformer, and $\mathbf{r}(t)$ is the array output vector described in Equation (2.5). The output average power of the beamformer may then be calculated as

$$E\{|r_B(t)|^2\} = \mathbf{w}^H\mathbf{R}\mathbf{w}. \tag{2.9}$$

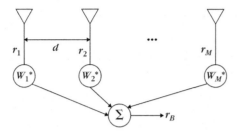

FIGURE 2.2

Beamformer.

Although the array output vector $\mathbf{r}(t)$ presented in Equation (2.4) was derived for a uniform linear array, this equation may apply to an arbitrary array geometry as long as the adequate array manifold matrix $\mathbf{A}$, obtained analytically or by calibration, is considered in the model. The array or beamformer outputs may be sampled at different time instants in order to obtain an ensemble of array snapshots that may be used in the estimation of the angles of arrival θ_i, $i = 1, 2, \ldots, D$. In the following sections we will discuss AOA estimation techniques commonly applied by position estimation schemes. In what follows, we will assume that the number of signal sources D is known or has been successfully estimated. Methods to estimate the number of sources may be found in [47, 48].

2.1.2 Cramer Rao Bound for Array Observations

The Cramer Rao bound (CRB) [38] obtains expressions for the best possible error covariance that may be obtained by any unbiased parameter estimator (independent of the particular estimator's form). The CRB for the observation model given in Equation (2.5) was found in [40] for the case of signals $\{s_i(t), i = 1, \ldots, D\}$ with known correlation matrix and unknown noise power σ^2. More general performance bounds may be found in [25, 46]. Given all the assumptions made in the previous section ($M > D$, nonsingular $\mathbf{S}$, linear independence between columns of $\mathbf{A}$, white Gaussian noise), the CRB for any unbiased estimator of the vector of AOAs $\boldsymbol{\theta} = [\theta_1 \ldots \theta_D]^T$, which is based on N independent snapshot observations of M sensor outputs at times $\{t_1 \ldots t_N\}$ is given by

$$\text{CRB}(\boldsymbol{\theta}) = \frac{\sigma}{2} \left\{ \sum_{l=1}^{N} Re \left[\mathbf{Y}(t_l)^H \mathbf{D}^H [\mathbf{I} - \mathbf{A}(\mathbf{A}^H \mathbf{A})^{-1} \mathbf{A}^H] \mathbf{D} \mathbf{Y}(t_l) \right] \right\}^{-1},$$

$$(2.10)$$

$$VAR_{CRB}(\sigma) = \frac{\sigma^2}{MN},$$

where $\mathbf{Y}(t) = \mathrm{diag}\{s_1(t), \ldots, s_D(t)\}$, and $\mathbf{D} = \left[\frac{d\mathbf{a}_1(\theta)}{d\theta} \cdots \frac{d\mathbf{a}_D(\theta)}{d\theta} \right]$. This bound on the error covariance may be used to compare the performance of the specific parameter estimators to be presented in the following discussion.

2.2 NONPARAMETRIC METHODS FOR AOA ESTIMATION

Nonparametric methods make no assumptions about the correlation structure of the data and assume that the array is calibrated—that is, that the form of the steering vector $\mathbf{a}(\theta)$ is known.

2.2.1 Beamscan AOA Estimator

Analogous to power spectral density estimators, a spatial periodogram may be defined [42] as

$$|\mathbf{a}(\theta)^H \mathbf{r}(t)|^2 = \left| \sum_{m=1}^{M} r_m(t) e^{j(m-1)k} \right|^2, \tag{2.11}$$

which may be computed using a fast Fourier transform (FFT). When the steering vector is scanned over θ, this periodogram yields large variance estimates of the spatial spectral density of the array output vector $\mathbf{r}(t)$. To reduce this variance, several periodograms obtained with N array output snapshots may be averaged to yield

$$P_{\mathrm{BS}}(\theta) = \frac{1}{N} \sum_{t=1}^{N} |\mathbf{a}(\theta)^H \mathbf{r}(t)|^2 = \mathbf{a}(\theta)^H \hat{\mathbf{R}} \mathbf{a}(\theta), \tag{2.12}$$

where

$$\hat{\mathbf{R}} = \frac{1}{N} \sum_{t=1}^{N} \mathbf{r}(t) \mathbf{r}(t)^H, \tag{2.13}$$

which corresponds to the N-snapshot estimate of the array correlation matrix. The estimates of $\theta_i, i = 1, 2, \ldots, D$ may be chosen as the D largest peaks of the beamscan estimator function $P_{\mathrm{BS}}(\theta)$.

The resolution of the beamscan estimator is inversely proportional to the array aperture and, for large array lengths and $d = \frac{\lambda}{2}$, it may be approximated as $\frac{1}{M}$. The beamscan estimator is consistent for the case of a single source. Multiple sources, however, will cause inconsistent AOA estimates, and their bias will grow when the sources are strongly correlated or closely spaced. Note, however, that this estimator will obtain accurate AOA estimates when the sources are uncorrelated and separated by more than the array beam width. Closely spaced and highly correlated sources may arise in dense multipath scenarios where multiple delayed copies of a signal may reach the array from different directions. These multiple signal copies may be caused by signal reflections and scattering from objects such as buildings, cars, trees, people, and so on. If the reflecting objects are close to each other, so are the AOAs of the different reflections.

2.2.2 MVDR AOA Estimator

MVDR stands for *minimum variance distortionless response*, and the estimator may be obtained by minimizing the average output power of a beamformer $E\{|y(t)|^2\}$ as given in Equation (2.9) subject to the linear constraint $\mathbf{w}^H \mathbf{a}(\theta) = 1$. It is easy to verify that the minimization is achieved by the weight vector [18],

$$\mathbf{w} = \frac{\mathbf{R}^{-1}\mathbf{a}(\theta)}{\mathbf{a}(\theta)^H \mathbf{R}^{-1} \mathbf{a}(\theta)}. \qquad (2.14)$$

The idea is to minimize the array response to every AOA, except a prescribed angle of interest θ. Inserting this optimal weight into Equation (2.9), and replacing the spatial correlation matrix $\mathbf{R}$ by its sample estimate counterpart $\hat{\mathbf{R}}$ yields the MVDR AOA estimator,

$$P_{\text{MVDR}}(\theta) = \frac{1}{\mathbf{a}(\theta)^H \hat{\mathbf{R}}^{-1} \mathbf{a}(\theta)}. \qquad (2.15)$$

The estimates of θ_i, $i = 1, 2, \ldots, D$ may be chosen as the D largest peaks of $P_{\text{MVDR}}(\theta)$ when "steered" over all possible values of θ. Note that the sample correlation matrix estimate $\hat{\mathbf{R}}$ will be nonsingular if the noise correlation matrix is nonsingular (which is clearly the case in Equation (2.7) where $\mathbf{R_n} = \sigma^2 \mathbf{I}$). The MVDR has been shown, empirically, to have better performance than the beamscan estimator [42].

2.3 PARAMETRIC METHODS FOR AOA ESTIMATION

Performance of the nonparametric AOA estimation techniques may be improved if prior information about the array model or the signal statistics is available.

2.3.1 Maximum Likelihood AOA Estimator

The maximum likelihood (ML) principle is based on the maximization of the likelihood function of the array observations, which follows from the joint density of N snapshots of the array outputs. Under the assumption of Gaussian uncorrelated noise, the joint probability density function for N array snapshots at times $\{t_l, l = 1 \ldots N\}$ is given by

$$f_{\mathbf{r}(t_1)\ldots\mathbf{r}(t_N)} = [(2\pi)^{-MN}(\sigma/2)^{MN}]\exp\left\{-\frac{1}{\sigma^2}\sum_{l=1}^{N}[\mathbf{r}(t_l) - \mathbf{As}(t_l)]^H[\mathbf{r}(t_l) - \mathbf{As}(t_l)]\right\}. \quad (2.16)$$

Taking the logarithm of this density and ignoring terms that do not depend on the AOA parameters θ_i, $i = 1 \ldots M$, we obtain the log-likelihood function (the negative sign will also be neglected with the effect of converting the maximization problem into a minimization one),

$$L = \frac{1}{\sigma^2}\sum_{l=1}^{N}[\mathbf{r}(t_l) - \mathbf{As}(t)]^H[\mathbf{r}(t_l) - \mathbf{As}(t)]. \quad (2.17)$$

Clearly, for the case of non-Gaussian noise, the above function corresponds to the sum of squared errors of the model whose minimization leads to a least-squares (LS) estimation scheme.

First we use the well-known LS solution [38, 42] to estimate the unknown signal amplitude vectors $\mathbf{s}(t_l)$ to obtain

$$\hat{\mathbf{s}}(t_l) = (\mathbf{A}^H\mathbf{A})^{-1}\mathbf{A}^H\mathbf{r}(t_l), \quad l = 1 \ldots N. \quad (2.18)$$

Inserting this result into Equation (2.18), we obtain

$$L = \frac{N}{\sigma^2}tr\left[\mathbf{I} - \mathbf{A}(\mathbf{A}^H\mathbf{A})^{-1}\mathbf{A}^H\hat{\mathbf{R}}\right], \quad (2.19)$$

where $\hat{\mathbf{R}}$ is the sample array correlation matrix given in Equation (2.13). It follows from this last equation that the ML estimate for θ_i, $i = 1, 2, \ldots, D$ is obtained by solving

$$\{\hat{\theta}_i\}_{i=1}^{D} = \arg\max_{\{\theta_i\}_{i=1}^{D}} tr\left[\mathbf{A}(\mathbf{A}^H\mathbf{A})^{-1}\mathbf{A}^H\hat{\mathbf{R}}\right], \quad (2.20)$$

which corresponds to a nonlinear multidimensional maximization problem. The solution to this type of problem will usually be computationally intensive and will suffer from local minima and convergence issues. Implementations of the AOA ML estimator have been discussed in several works (see references 10-12, 33-36, 42 in [40]).

According to Stoica and Nehorai [40], the ML estimator is inefficient in the sense that an infinite number of snapshots will not be enough for the estimator to reach the CRB as long as the number of sensors in the array M is finite. The CRB will only be reached by letting $M \rightarrow \infty$, which could only be achieved by an infinite aperture array.

An interesting observation in Stoica and Moses [42] states that the beamscan AOA estimator provides an approximate solution to the ML problem whenever the AOAs are well separated. In this scenario, it can be seen that $\mathbf{A}^H\mathbf{A} \approx M\mathbf{I}$. Then the objective function in Equation (2.20) may be approximated as

$$tr[\mathbf{A}(\mathbf{A}^H\mathbf{A})^{-1}\mathbf{A}^H\hat{\mathbf{R}}] \approx \frac{1}{M}\sum_{i=1}^{D}\mathbf{a}(\theta_i)^H\hat{\mathbf{R}}\mathbf{a}(\theta_i), \qquad (2.21)$$

where the right side of the equation will be maximized at the true AOAs as was previously explained in the beamscan estimator discussion.

Due to the complexity of the AOA ML estimator, several other suboptimal estimators are usually preferred, and actually many of these suboptimal schemes will reach an optimal performance under specific scenarios [40, 41].

2.3.2 MUSIC Algorithm for AOA Estimation

The multiple signal classification (MUSIC) algorithm [39] falls into the category of the so-called subspace, or super-resolution, estimators. These algorithms profit from the eigenstructure properties of the array correlation matrix to obtain very-high-resolution estimators with lower computational complexity when compared to ML estimation schemes.

Let us analyze the properties of the spatial correlation matrix $\mathbf{R}$ described in Equation (2.7). It is clear that if the number of array sensors is larger than the number of signal sources (i.e., $M > D$), when $\mathbf{S}$ is positive definite (i.e., the signals $\mathbf{s}_i(t)$ are not fully correlated), the matrix $\mathbf{R} - \mathbf{R_n}$ will have rank D and a null space of dimension $M - D$. Then matrix $\mathbf{R}$ will have D eigenvalues greater than σ^2 and $M - D$ eigenvalues equal to σ^2; these eigenvalues may be ordered from largest to smallest such that $\lambda_1 > \lambda_2 > \cdots > \lambda_D > \lambda_{D+1} = \lambda_{D+2} = \ldots = \lambda_M = \sigma^2$. The eigenvectors $\{\mathbf{e}_1, \mathbf{e}_2, \ldots, \mathbf{e}_D\}$ corresponding to the largest eigenvalues span the D-dimensional *signal subspace*. We can group these signal eigenvectors in the columns of matrix $\mathbf{E}_s$. The eigenvectors $\{\mathbf{e}_{D+1}, \mathbf{e}_{D+2}, \ldots, \mathbf{e}_M\}$ corresponding to the smallest eigenvalues span the $(M - D)$-dimensional *noise subspace*. These signal eigenvectors can be grouped in the columns of matrix $\mathbf{E}_n$. If we write the eigenvalue problem,

$$\mathbf{R}\mathbf{e}_i = \lambda_i\mathbf{e}_i; \quad i = D+1, D+2, \ldots, M, \qquad (2.22)$$

we may substitute $\mathbf{R}$ by the leftmost side of Equation (2.7), and realize that the eigenvalue λ_i for every noise subspace eigenvector is equal to σ^2 to obtain

$$\mathbf{A}\mathbf{S}\mathbf{A}^H\mathbf{e}_i + \sigma^2\mathbf{e}_i - \sigma^2\mathbf{e}_i = \mathbf{0}; \quad i = D+1, D+2, \ldots, M, \qquad (2.23)$$

which implies that

$$\mathbf{A}^H\mathbf{e}_i = \mathbf{0}; \quad i = D+1, D+2, \ldots, M. \qquad (2.24)$$

Then it is clear that the columns of matrix $\mathbf{A}$ are orthogonal to the eigenvectors that span the noise subspace.

For an exactly known $\mathbf{R}$, we can find the desired angles $\theta_i, i = 1, 2, \ldots, D$, by the following steps that comprise the MUSIC algorithm.

1. Compute the eigenvalues and eigenvectors of the $(M \times M)$ matrix $\mathbf{R}$.

2. D and σ^2 are determined by the facts that the minimum eigenvalue is equal to σ^2 and has multiplicity $M - D$.

3. Group the spanning eigenvectors for the noise subspace in the set $\{\mathbf{e}_{D+1}, \mathbf{e}_{D+2}, \ldots, \mathbf{e}_M\}$.

4. Perform an orthogonality test between different candidate steering vectors and the noise subspace to obtain the MUSIC pseudospectrum,

$$P_{\text{MUSIC}}(\theta) = \frac{\mathbf{a}(\theta)^H \mathbf{a}(\theta)}{\sum_{j=D+1}^{M} [\mathbf{a}(\theta)^H \mathbf{e}_j]^2}. \tag{2.25}$$

Ideally $P_{\text{MUSIC}}(\theta)$ will peak to infinity each time a true $\theta_i, i = 1, 2, \ldots, D$ angle is tested. When neither D nor $\mathbf{R}$ is known, they may be estimated using *model-order identification* techniques presented in *MDL* or *AIC* [47, 48], and an N-snapshot sample covariance matrix, respectively.

We have seen that MUSIC works only when the rank of matrix $\mathbf{R} - \mathbf{R_n}$ is equal to D, that is, when the signals are uncorrelated. Unfortunately, in many multipath scenarios of interest, should the signals incident on the array be strongly or completely correlated, the rank of this matrix would reduce to unity. In this case we will not be able to find the correct noise subspace and the MUSIC algorithm will fail. In order to restore the rank of the matrix $\mathbf{R} - \mathbf{R_n}$, a well-known array processing technique known as *spatial smoothing* may be used [33, 34, 37].

A closed-form expression for the MUSIC AOA asymptotic estimation variance (large N) is given by

$$VAR_{\text{MUSIC}}(\hat{\theta}_i) = \frac{\sigma}{2N} \left\{ [\mathbf{S}^{-1}]_{ii} + \sigma[\mathbf{S}^{-1}(\mathbf{A}^H \mathbf{A})^{-1}\mathbf{S}^{-1}]_{ii} \right\} / h(\theta_i),$$
$$h(\theta) = \mathbf{a}'(\theta)^H [\mathbf{I} - \mathbf{A}(\mathbf{A}^H \mathbf{A})^{-1}\mathbf{A}^H]\mathbf{a}'(\theta), \tag{2.26}$$

where $\mathbf{a}'(\theta) = \frac{d\mathbf{a}(\theta)}{d\theta}$ [40]. Apart from the obvious result that variance increases with the noise power, this equation also shows that variance will increase if the signals are highly correlated (since that would imply that $\mathbf{S}$ is nearly singular), or if their AOAs are closely spaced (which would imply that $\mathbf{A}^H \mathbf{A}$ is nearly singular). Also, Xu et. al. [8, 49] presented results that show that MUSIC may suffer from large bias in scenarios with low SNR, closely spaced sources, or highly correlated sources.

For uncorrelated signals, MUSIC reaches the CRB for large N, M, and SNR scenarios. On the other hand, correlated signals prevent MUSIC from achieving the CRB and may actually cause it to be inefficient even in the presence of large N, M, and SNR cases. Finally, MUSIC is very sensitive to array calibration errors, as has been observed in several works (e.g., [15, 44]).

2.3.3 ESPRIT Algorithm for AOA Estimation

ESPRIT (estimation of signal parameters via rotational invariance techniques) [36] eliminates the array calibration required by MUSIC and offers some computational advantages. It exploits the rotational invariance in the signal subspace that is created by two arrays with a translational invariant structure.

Consider an array with $M + 1$ sensors and two M-dimensional array data vectors $\mathbf{r}_L(t) = [r_1(t) \ldots r_M(t)]^T$ and $\mathbf{r}_U(t) = [r_2(t) \ldots r_{M+1}(t)]^T$, where we recall that $r_m(t)$ is the m-th sensor signal output as given in Equation (2.2). Each of the array data vectors may be expressed as

$$\mathbf{r}_L(t) = \mathbf{A}\mathbf{s}(t) + \boldsymbol{\eta}_L(t),$$
$$\mathbf{r}_U(t) = \mathbf{A}\boldsymbol{\Phi}\mathbf{s}(t) + \boldsymbol{\eta}_U(t),$$

(2.27)

where $\boldsymbol{\Phi} = \text{diag}\{e^{jk_1}, e^{jk_2}, \ldots, e^{jk_D}\}$ (recall that $k_i = \frac{\omega_c d}{u} \sin \theta_i$ as was defined in Equation (2.3)). Collecting both array responses in a single vector, we obtain

$$\bar{\mathbf{r}}(t) = \begin{bmatrix} \mathbf{r}_L(t) \\ \mathbf{r}_U(t) \end{bmatrix} = \begin{bmatrix} \mathbf{A} \\ \mathbf{A}\boldsymbol{\Phi} \end{bmatrix} \mathbf{s}(t) + \begin{bmatrix} \boldsymbol{\eta}_L(t) \\ \boldsymbol{\eta}_U(t) \end{bmatrix},$$

(2.28)

$$\bar{\mathbf{r}}(t) = \bar{\mathbf{A}}\mathbf{s}(t) + \bar{\boldsymbol{\eta}}(t).$$

The objective is to estimate the elements of $\boldsymbol{\Phi}$ that contain the AOA information without the need to know the array manifold $\mathbf{A}$.

The spatial correlation matrices for $\mathbf{r}_L(t)$, $\mathbf{r}_U(t)$, and $\bar{\mathbf{r}}$ are given, respectively, by

$$\mathbf{R}_L = E\{\mathbf{r}_L(t)\mathbf{r}_L(t)^H\} = \mathbf{A}\mathbf{S}\mathbf{A}^H + \sigma^2\mathbf{I},$$
$$\mathbf{R}_U = E\{\mathbf{r}_U(t)\mathbf{r}_U(t)^H\} = \mathbf{A}\boldsymbol{\Phi}\mathbf{S}\boldsymbol{\Phi}^H\mathbf{A}^H + \sigma^2\mathbf{I},$$

(2.29)

$$\bar{\mathbf{R}} = E\{\bar{\mathbf{r}}(t)\bar{\mathbf{r}}^H(t)\} = \bar{\mathbf{A}}\mathbf{S}\bar{\mathbf{A}}^H + \sigma^2\boldsymbol{\Sigma},$$

where the normalized noise covariance matrix $\boldsymbol{\Sigma}$ is given by

$$\boldsymbol{\Sigma} = \begin{bmatrix} \mathbf{I} & \mathbf{Q}_{-1} \\ \mathbf{Q}_{+1} & \mathbf{I} \end{bmatrix},$$

(2.30)

and $\mathbf{Q}_{-1}$ and $\mathbf{Q}_{+1}$ are $M \times M$ matrices with ones immediately below the main diagonal and zeros elsewhere, and ones immediately above the main diagonal and zeros elsewhere, respectively. The identity matrix $\mathbf{I}$ is also an $M \times M$ matrix. The D largest eigenvectors of matrices $\mathbf{R}_L$ and $\mathbf{R}_U$ span the signal subspace for each subarray and may be grouped in the columns of matrices $\mathbf{E}_{s,L}$ and $\mathbf{E}_{s,U}$, respectively. Further, solving the generalized eigenvalue problem for matrix $\bar{\mathbf{R}}$ given by

$$\bar{\mathbf{R}}\bar{\mathbf{e}}_i = \bar{\lambda}_i \mathbf{\Sigma} \bar{\mathbf{e}}_i \qquad (2.31)$$

will yield $2M - D$ smallest generalized eigenvalues equal to σ^2 and D generalized eigenvalues greater than σ^2. The corresponding set of largest generalized eigenvectors may be grouped in the columns of matrix $\bar{\mathbf{E}}_s$ that span the signal subspace for the entire array. Due to the invariance structure of the array, $\bar{\mathbf{E}}_s$ may be decomposed into the signal subspaces $\mathbf{E}_{s,L}$ and $\mathbf{E}_{s,U}$. Since the arrays are translationally related, there should exist a unique nonsingular transformation matrix $\mathbf{\Psi}$ such that

$$\mathbf{E}_{s,L}\mathbf{\Psi} = \mathbf{E}_{s,U}. \qquad (2.32)$$

Similarly, a nonsingular transformation matrix $\mathbf{T}$ exists such that

$$\bar{\mathbf{E}}_s = \bar{\mathbf{A}}\mathbf{T} = \begin{bmatrix} \mathbf{E}_{s,L} \\ \mathbf{E}_{s,U} \end{bmatrix} = \begin{bmatrix} \mathbf{A}\mathbf{T} \\ \mathbf{A}\mathbf{\Phi}\mathbf{T} \end{bmatrix}. \qquad (2.33)$$

It follows from Equations (2.32) and (2.33) that

$$\mathbf{T}\mathbf{\Psi}\mathbf{T}^{-1} = \mathbf{\Phi}. \qquad (2.34)$$

Clearly, Equation (2.34) has the form of an eigenvalue problem where the columns of matrix $\mathbf{T}$ collect the eigenvectors of matrix $\mathbf{\Psi}$, and the diagonal terms of matrix $\mathbf{\Phi}$ collect the corresponding eigenvalues. Hence, estimation of the transformation matrix $\mathbf{\Psi}$ and its corresponding eigenvalues will yield the AOA estimates embedded in the diagonal terms of $\mathbf{\Phi}$.

A total least-squares (TLS) criterion is usually applied for the estimation of matrix $\mathbf{\Psi}$ [16, 36]. The steps of the TLS-based ESPRIT follow:

1. Obtain an estimate of the total array output correlation matrix $\hat{\bar{\mathbf{R}}}$ from N snapshot observations of the array outputs.

2. Compute the generalized eigenvalue problem,

$$\hat{\bar{\mathbf{R}}}\bar{\mathbf{E}} = \mathbf{\Sigma}\bar{\mathbf{E}}\mathbf{\Lambda}, \qquad (2.35)$$

where $\mathbf{\Lambda} = \text{diag}\{\lambda_1, \ldots, \lambda_{2M}\}, \lambda_1 \geqslant \ldots \geqslant \lambda_{2M}$, and $\bar{\mathbf{E}} = [\mathbf{e}_1 \ldots \mathbf{e}_{2M}]$ (i.e., the set of all generalized eigenvectors ordered from the largest to the smallest).

3. Estimate the number of sources as the D largest generalized eigenvalues in $\mathbf{\Lambda}$ or with the model order estimation techniques that have been mentioned in previous sections.

4. Obtain the signal subspace matrix $\bar{\mathbf{E}}_s$ (from the D largest generalized eigenvectors) and decompose it to obtain $\mathbf{E}_{s,L}$ and $\mathbf{E}_{s,U}$ as in Equation (2.33).

5. Form a new matrix,

$$\mathbf{C} = \begin{bmatrix} \mathbf{E}_{s,L}^H \\ \mathbf{E}_{s,U}^H \end{bmatrix} [\mathbf{E}_{s,L} \, \mathbf{E}_{s,U}], \tag{2.36}$$

and perform its eigendecomposition to obtain a set of eigenvalues $\lambda_{c,1} \geqslant \ldots \geqslant \lambda_{c,2D}$ and a corresponding eigenvector set collected in columns of matrix $\mathbf{E}_c$.

6. Partition $\mathbf{E}_c$ such that

$$\mathbf{E}_c = \begin{bmatrix} \mathbf{E}_{11} & \mathbf{E}_{12} \\ \mathbf{E}_{21} & \mathbf{E}_{22} \end{bmatrix}. \tag{2.37}$$

7. Estimate the rotation operator $\mathbf{\Psi}$ as

$$\mathbf{\Psi} = -\mathbf{E}_{12}\mathbf{E}_{22}^{-1}, \tag{2.38}$$

and calculate its eigenvalues $\{\lambda_{\Psi,1} \ldots \lambda_{\Psi,D}\}$.

8. Finally, estimate the AOAs, noting that $\lambda_{\Psi,i} = |\lambda_{\Psi,i}| e^{j \arg(\lambda_{\Psi,i})}$, and then

$$\theta_i = \sin^{-1}\left[\frac{u}{\omega_c d} \arg(\lambda_{\Psi,i})\right]. \tag{2.39}$$

As discussed in Ottersten et al. [25], the ESPRIT and MUSIC algorithms have comparable performance under most scenarios of interest with the advantage that the former reaches its asymptotic behavior with far fewer snapshots, is robust to array calibration errors, and is computationally simpler since it does not require a search over θ to find AOA estimates.

2.4 TOA AND TDOA MEASUREMENTS

First time of arrival may be directly related to the "time of flight" of a transmitted signal. If the propagation speed of the transmission medium is known, and if a direct line of sight (LOS) exists, then the distance between the transmitter and receiver may be directly related to this time of flight.

The basic problem of TOA-based localization techniques is to estimate the propagation delay of a signal arriving from the direct LOS propagation path.

TDOA measurements may be estimated directly from observations at two separate sensors, or indirectly through two TOA estimates obtained at each sensor. The former requires synchronization between the receivers while the latter requires synchronization between the receivers and the transmitter.

An important property of TOA/TDOA-based ranging techniques is that their variance does not increase with distance as is the case in AOA and RSS-based schemes.

2.4.1 The Time of Arrival Problem

Consider a known signal $s(t)$ transmitted through a single path channel. In this scenario, the received signal may be modeled as

$$r(t) = m \cdot s(t - \tau) + \eta(t), \tag{2.40}$$

where m is a random complex-valued gain, τ is the signal propagation delay, and $\eta(t)$ is a stationary zero mean noise process. The ML estimate of τ can be obtained by finding the value of ζ that maximizes the cross-correlation function between the received and transmitted signals when estimated over an observation window of T seconds [30],

$$\hat{p}_{rs}(\zeta) = \frac{1}{T} \int_0^T r(t)s(t - \zeta)dt = m\hat{p}_{ss}(\zeta - \tau) + v(\zeta), \tag{2.41}$$

where ζ is the lag time, $\hat{p}_{ss}(\zeta)$ is the finite sample estimate of the autocorrelation function of the transmitted signal, and $v(\zeta)$ is the noise as seen at the output of the correlator. Note that we have used the notation $\hat{x}$ to denote the time average estimate of the corresponding exact statistical moment x. This notation will prove useful in Section 2.4.5 when we introduce the generalized correlator. Then, even though $s(t)$ and its exact autocorrelation function are known, the form of this autocorrelation may differ from the exact one at the output of the correlator in Equation (2.41) when T is not large enough, and for this reason we refer to this finite sample estimate as $\hat{p}_{ss}(\zeta)$.

Now consider a known signal $s(t)$ that is transmitted through a multipath channel with impulse response given by

$$h(t) = \sum_{i=1}^{D} m_i \delta(t - \tau_i), \tag{2.42}$$

where D is the number of existing paths, $\{\tau_i ; i = 1, \dots, D\}$ are the time-delay parameters or the times of arrival of the echoes that include the first time of arrival τ_1, which we want to estimate, and $\{m_i ; i = 1, \dots, D\}$ are random complex-valued gains that include the scattering and propagation fading effects. In a period of time starting at the moment the signal is transmitted and ending at a time when it is assumed that no more echoes will arrive, a sensor receives a superposition of delayed and scaled versions of $s(t)$ and the received-signal baseband model can be written as

$$r(t) = \sum_{i=1}^{D} m_i s(t - \tau_i) + \eta(t), \tag{2.43}$$

where $\eta(t)$ is an additive white-noise process with a spectral level of σ^2.

The previously described ML cross-correlation scheme may be applied to the received signal to obtain

$$\hat{p}_{rs}(\zeta) = \sum_{i=1}^{D} m_i \hat{p}_{ss}(\zeta - \tau_i) + v(\zeta), \qquad (2.44)$$

which corresponds to a weighted sum of multiple shifted-signal autocorrelation functions. The correct estimation of the first time of arrival in the multipath scenario will strongly depend on the instantaneous channel response at the time of the measurements, and the signal bandwidth. The performance of the ML estimator presented in Equation (2.44) will degrade considerably whenever the first TOA has a strong fade (due to shadowing or small-scale fading) or when the delay differences between adjacent paths and the first TOA are smaller than the duration of the main signal's autocorrelation peak. The solution to these problems may be achieved by increasing the signal bandwidth (which in turn will decrease the autocorrelation peak duration), or by applying other high-resolution signal processing techniques that will be described shortly.

2.4.2 The Time Difference of Arrival Problem

In the TDOA problem, a signal $s(t)$ transmitted at a remote location is received by two spatially separated receivers. For the case of a flat fading channel, the signals at the two receivers may be described as

$$\begin{aligned} r_a(t) &= s(t) + \eta_a(t), \\ r_b(t) &= ms(t + \tau) + \eta_b(t), \end{aligned} \qquad (2.45)$$

where again m is a random complex-valued gain, and now τ corresponds to the time difference with which the signal arrived at each of the two receivers. It is assumed that $\eta_a(t)$ and $\eta_b(t)$ are stationary noise processes. A correlation procedure similar to the one presented for the TOA case may be implemented to estimate τ [23] as follows:

$$\hat{p}_{r_a r_b}(\zeta) = \frac{1}{T} \int_0^T r_a(t) r_b(t - \tau) dt = m\hat{p}_{ss}(\zeta - \tau) + v(\zeta), \qquad (2.46)$$

where again $\hat{p}_{ss}(\zeta)$ is the finite sample estimate of the autocorrelation of $s(t)$, and $v(\zeta)$ includes the noise terms as seen at the output of the correlator. Then the TDOA may be estimated by finding the value of τ that maximizes Equation (2.46).

In the multipath channel case, the signals received at each sensor may be expressed as

$$r_a(t) = \sum_{i=1}^{D} m_{a,i} s(t - \tau_{a,i}) + \eta_a(t),$$

$$r_b(t) = \sum_{i=1}^{D} m_{b,i} s(t - \tau_{b,i}) + \eta_b(t),$$

(2.47)

where subindexes a and b are used to distinguish the parameters corresponding to each of the multipath channels (corresponding to the signal travel from the transmitter to each of the a and b receivers). The estimated cross-correlation function of the received signals is calculated as

$$\hat{p}_{r_a r_b}(\zeta) = \sum_{i=0}^{D} \sum_{j=0}^{D} m_{a,i} m_{b,j} \hat{p}_{ss}(\zeta - (\tau_{a,i} - \tau_{b,i})) + v(\zeta).$$

(2.48)

The TDOA estimate will consist of the difference $\tau_{a,1} - \tau_{b,1}$. As stated in Li et al. [23], it is clear in this scenario that the TDOA cannot be detected by finding the delay value that corresponds to the first peak of the cross-correlation function since it will not necessarily correspond to the delay value $\tau_{a,1} - \tau_{b,1}$. Further, the strongest peak of the cross-correlation will not necessarily occur at $\tau_{a,1} - \tau_{b,1}$. Then estimation of the TDOA becomes quite difficult in the presence of multipath since the correlation peak at the true TDOA is not guaranteed to be the first one or the strongest. For this reason, in many applications where the transmitted signal is known a priori, TDOA is measured based on two separate TOA measurements obtained at each receiver. This scheme avoids measurement ambiguities but requires synchronization between the transmitter and both of the receivers. Direct TDOA measurements, on the other hand, only require synchronization between the receivers.

2.4.3 Performance Bound for TOA and TDOA Problems

Lower bounds for the estimation variance of delay in the neighborhood of the true time delay were described in Quazi [30] (and references therein) for the single-path propagation channel model described in Equations (2.40) and (2.45). The derivations assume that the signal and noise spectral densities are constant over the bandwidth $W = f_2 - f_1$ Hz with power of P_0 and σ_n^2 respectively. The bounds are given by

$$\sigma_\tau \geq \frac{1}{\sqrt{8\pi^2}} \frac{1}{SNR} \frac{1}{TW} \frac{1}{f_c} \frac{1}{\sqrt{1 + \frac{W^2}{12f_c^2}}}, \quad SNR \ll 1,$$

$$\sigma_\tau \geq \frac{1}{\sqrt{2\pi^2}} \frac{1}{\sqrt{SNR}} \frac{1}{TW} \frac{1}{f_c} \frac{1}{\sqrt{1 + \frac{W^2}{12f_c^2}}}, \quad SNR \gg 1.$$

(2.49)

The bounds in Equation (2.49) show that TOA estimation variance is inversely proportional to $SNR = P_0/\sigma^2$, bandwidth time (BT) product TW, and center carrier frequency f_c.

Lower bounds for the variance of estimates of delay in the presence of multipath have been presented in [4, 5], where it has been shown that first TOA estimation accuracy is strongly dependent on delay separation and relative phase between the first path and its closer clutter of arrivals.

2.4.4 Received Signal Model and Its Analogy to the Array Processing Problem

Consider repeating the transmission of a signal $s(t)$ every T seconds through a D-path channel, such as the one described by the impulse response in Equation (2.42), to obtain N responses (referred to as *snapshots*). We assume that after T seconds, no more echoes arrive so that each snapshot has no interference from previous transmissions. It is also assumed that the path delay locations $\{\tau_i;\ i=1,\ldots,D\}$ do not change during the N-snapshot acquisition period. Then, from Equation (2.43), each snapshot may be expressed as

$$r_j(t) = \sum_{i=1}^{D} m_{ij}s(t-\tau_i) + \eta_j(t) \quad j=1,2,\ldots,N, \tag{2.50}$$

where m_{ij} stands for the complex gain of the i-th path as observed at the time when the j-th snapshot was captured.

The TOA estimation problem can be stated as follows: Given a set of snapshots as above, and assuming that we know the number of echoes D (they can be estimated using the MDL or AIC methods [47, 48]), find the time delays corresponding to each return. Note that contrary to the AOA problem where we had multiple sensors, observations for the TOA problem consist of a sequence of N snapshots from a *single sensor*. The analogy to the M sensors is introduced if we sample the received waveforms $r_j(t)$ at M instants and collect the successive samples in a vector of length M. Then for each snapshot we can write in vectorial notation [7]:

$$\begin{bmatrix} r_j(t_1) \\ r_j(t_2) \\ \vdots \\ r_j(t_M) \end{bmatrix} = [\mathbf{a}(\tau_1)\,\mathbf{a}(\tau_2)\cdots\mathbf{a}(\tau_D)] \begin{bmatrix} m_{1j} \\ m_{2j} \\ \vdots \\ m_{Dj} \end{bmatrix} + \begin{bmatrix} \eta_j(t_1) \\ \eta_j(t_2) \\ \vdots \\ \eta_j(t_M) \end{bmatrix}, \tag{2.51}$$

$$\mathbf{r}_j = \mathbf{A}\mathbf{g}_j + \eta_j, \tag{2.52}$$

where now the "steering vectors" are defined as

$$\mathbf{a}(\tau) = [s(t_1-\tau)\,s(t_2-\tau)\,\ldots\,s(t_M-\tau)]^T, \tag{2.53}$$

and the correlation matrix is given by

$$R = E\{r_j r_j^H\} = AGA^H + R_n, \tag{2.54}$$

where $A = [a(\tau_1) a(\tau_2) \ldots a(\tau_D)]$ is the "array manifold" matrix, and G and $R_n = \sigma^2 I$ are the gain and noise correlation matrices, respectively. For each τ, the steering vector can be calculated in a simple way since we know the signal form. In this formulation it is clear that the rank of the matrix $R - R_n$ will be equal to D, provided that $M > D$ and that the signal echoes are uncorrelated. For instance, if the channel path gains $\{m_i; i = 1, \ldots, D\}$ are random and change rapidly and independently with time, the path decorrelation assumption will hold. It is clear that Equations (2.5) and (2.52) and (2.7) and (2.54) have the same structure.

So the AOA estimation algorithms presented in Section 2.1 may be readily applied to the *one-sensor time delay estimation problem*, provided that the received signal correlation matrix and "steering vectors" are replaced by those given in Equations (2.54) and (2.53), respectively. Apart from the techniques presented in Section 2.1, there are other algorithms that profit from the time delay information obtained with the correlation function between a signal and its delayed counterpart. In the following we will discuss several of these algorithms; for instance, we will present the conventional pseudo-noise (PN) correlation method and its super-resolution counterpart implemented with MUSIC. We will also discuss TOA estimation techniques based on generalized cross-correlation and successive cancellation.

2.4.5 Generalized Cross-Correlation Method for TOA or TDOA Estimation

Let us return to the received signal model presented in Equation (2.45). As mentioned previously, an optimum delay estimator consists of a cross-correlator given by Equation (2.46). Note, however, that if the observation window length is limited (i.e., T may not be arbitrarily large), then the cross-correlation function estimator may become inaccurate, especially in low SNR scenarios. The length of the observation window may be limited for various reasons, and usually the most significant relates to the coherence time of the channel in which the signal and noise may be assumed to be stationary. Noise and finite observation times are then a major source of error in the cross-correlation estimator.

Another important limitation of the estimator may be related to signal bandwidth. A small bandwidth will yield an autocorrelation function that is broad around its peak at the zero lag. These broad autocorrelation peaks will overlap in multipath scenarios to yield new peaks at incorrect delay locations. To improve performance under the previously described scenarios, a *generalized*

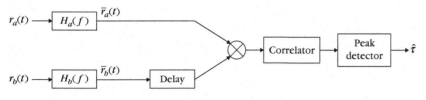

FIGURE 2.3

Generalized correlator.

correlation (GC) method may be implemented instead of the simple correlator. The results presented in this section were obtained from the works of Knapp and Carter [19] and Hassab and Boucher [17]. The GC may be implemented as shown in Figure 2.3 [19]. Prior to the cross-correlation operation, received signals $r_a(t)$ and $r_b(t)$ may be filtered through $H_a(f)$ and $H_b(f)$, respectively. The resulting signals $\bar{r}_a, \bar{r}_b$ are multiplied, integrated, and squared for a range of time shifts ζ, until a peak is obtained at a specific delay and is chosen as the estimate of τ. Note that when the filter's transfer functions are $H_a(f) = H_b(f) = 1$, we return to the simple cross-correlation form in Equation (2.46). The cross-correlation between the received signals may be related to the cross-power spectral density $G_{r_a r_b}(f)$ by [29],

$$p_{r_a r_b}(\zeta) = \int_{-\infty}^{\infty} G_{r_a r_b}(f) e^{j2\pi f \zeta} df. \tag{2.55}$$

After the signals have been filtered, the cross-power spectrum becomes

$$\bar{G}_{r_a r_b}(f) = H_a(f) H_b(f)^* G_{r_a r_b}(f). \tag{2.56}$$

Then the GC between $r_a(t)$ and $r_b(t)$ is given by

$$\bar{p}_{\bar{r}_a \bar{r}_b}(\zeta) = \int_{-\infty}^{\infty} \psi(f) G_{r_a r_b}(f) e^{j2\pi f \zeta} df, \tag{2.57}$$

where $\psi(f) = H_a(f) H_b(f)^*$. Note that in practice, the cross-spectral density between signals $r_a(t)$ and $r_b(t)$ is not known and needs to be estimated from a finite number of observations.

Let us explain the motivation for the use of a GC for estimation of delay. Without making any assumptions about the signal and noise statistics, the statistical cross-correlation between signals $r_a(t)$ and $r_b(t)$ is calculated as

$$p_{r_a r_b}(\zeta) = E\{r_a(t) r_b(t - \zeta)\} = m \cdot p_{ss}(\zeta - \tau) + E\{\eta_a(t)\eta_b(t - \zeta)\} + \phi_s(\zeta), \tag{2.58}$$

$$p_{r_a r_b}(\zeta) = [m \cdot p_{ss}(\zeta) \otimes \delta(\zeta - \tau)] + p_{\eta_a \eta_b}(\zeta) + \phi_s(\zeta),$$

where $\phi_s(\zeta) = E\{s(t)\eta_b(t - \zeta) + m \cdot s(t + \tau - \zeta)\eta_b(t)\}$. Let us analyze the terms on the right side of the second row of Equation (2.58). The first term corresponds to an impulse at the true delay τ, which is being spread by the signal's auto-correlation function. Due to the fact that $p_{ss}(\zeta) < p_{ss}(0)$ [29], this term will in fact peak at the true delay but this peak will be broadened according to the form of the signal auto-correlation function. This broadening will become a serious problem in dense multipath scenarios (where paths will be closely spaced in time) because multiple broad peaks will overlap and cause peaks at incorrect delay locations. Finally, note that if the signal is white, then it will not cause any spreading. The second and third terms are due to the noise processes $\eta_a(t)$ and $\eta_b(t)$. If these terms become large, they may cause large peaks (larger than the signal peak) in $p_{r_a r_b}(\zeta)$, which will yield erroneous estimates of τ.

Note that in the ideal case where the noise processes are uncorrelated and independent from signal $s(t)$, these terms will become zero. Note, however, that even in this ideal scenario, the time smearing of the impulse at delay τ due to the form of the signal's autocorrelation function will not disappear. Thus, one motivation for the use of a GC is that one may chose a frequency weighting factor $\psi(f)$ to maximize the amplitude and "sharpness" of the generalized cross-correlation peaks while maintaining low estimation variances. The other motivation arises when the noise terms in Equation (2.58) are not equal to zero. Non-zero noise terms may arise, even when the uncorrelated noise processes assumption holds, when the exact statistical averaging of Equation (2.58) is replaced by finite sample time averaging (which will always be the case in "real-world" applications). Obviously, noise terms will also appear when $\eta_a(t)$ and $\eta_b(t)$ are correlated. In the presence of noise terms, the frequency weighting factor $\psi(f)$ may be chosen to filter out their spectral components from the cross-spectral density $G_{r_a r_b}(f)$ and thus minimize the occurrence of false peaks.

Several weighting functions have been presented in the literature [1, 2, 17, 19], and some are listed in Table 2.1. To give insight about the rationale of the GC estimator, we will discuss two methods: Roth [35] and phase transform (PHAT) [19]. The reader may consult [19] and references therein for more detail on other methods. In practice, the cross-spectral density $G_{r_a r_b}(f)$ will need to be replaced by its corresponding estimate.

The Roth Generalized Cross-Correlator

As shown in Table 2.1, Roth's GC sets the frequency-weighting function to $\psi(f) = \frac{1}{G_{r_a r_a}(f)}$. Substitution of this function in Equation (2.57) yields

$$\hat{p}_{\tilde{r}_a r_b}(\zeta) = \int_{-\infty}^{\infty} \frac{\hat{G}_{r_a r_b}(f)}{G_{r_a r_a}(f)} e^{j2\pi f \zeta} df. \tag{2.59}$$

Table 2.1 Frequency Weighting for $\psi(f)$ for Generalized Cross-Correlation

Method	$\psi(f)$
Cross-Correlation	1
Roth	$\dfrac{1}{G_{r_a r_a}(f)}$
SCOT	$\dfrac{1}{[G_{r_a r_a}(f)G_{r_b r_b}(f)]^{1/2}}$
PHAT	$\dfrac{1}{|G_{r_a r_b}(f)|}$
Eckart	$\dfrac{G_{ss}(f)}{G_{\eta_a \eta_a}(f)G_{\eta_b \eta_b}(f)}$
ML	$\dfrac{|\gamma_{ab}(f)|^2}{|G_{r_a r_b}(f)|[1-\gamma_{ab}(f)|]^2}$

$$\gamma_{ab}(f)=\frac{G_{r_a r_b}(f)}{[G_{r_a r_a}(f)G_{r_b r_b}(f)]^{1/2}}$$

This last equation is an estimator of the optimum Wiener-Hopf filter, which best approximates the mapping of $r_a(t)$ to $r_b(t)$. In the presence of noise

$$G_{r_a r_a}(f)=G_{ss}(f)+G_{\eta_a \eta_a}(f). \tag{2.60}$$

Substituting this last equation, and Equation (2.58) into (2.59), and assuming uncorrelated noise processes η_a and η_b, and $\phi_s(\zeta)=0$, we obtain

$$\bar{p}_{\bar{r}_a \bar{r}_b}(\zeta)=\delta(\zeta-\tau)\otimes\int_{-\infty}^{\infty}\frac{mG_{ss}(f)}{G_{ss}(f)+G_{\eta_a \eta_a}(f)}e^{j2\pi f\zeta}df. \tag{2.61}$$

Note that the delta function will be broadened except for the case when $G_{\eta_a \eta_a}(f)=\text{const}\,G_{ss}(f)$. The Roth GC has the effect of suppressing frequency regions where $G_{\eta_a \eta_a}(f)$ is large, and where the cross-spectral density estimate $\hat{G}_{r_a r_b}(f)$ is more likely to be in error.

The Phase Transform Generalized Cross-Correlator

The PHAT frequency weighting function is given by $\psi(f)=\frac{1}{|G_{r_a r_b}(f)|}$. This yields the following GC:

$$\hat{\bar{p}}_{\bar{r}_a \bar{r}_b}(\zeta)=\int_{-\infty}^{\infty}\frac{\hat{G}_{r_a r_b}(f)}{|G_{r_a r_b}(f)|}e^{j2\pi f\zeta}df. \tag{2.62}$$

Then, again assuming uncorrelated noise processes η_a and η_b, and $\phi_s(\zeta)=0$, we find that

$$|G_{r_a r_b}(f)|=\alpha G_{ss}(f), \tag{2.63}$$

and for the case of perfect estimation of the cross-spectral density, $G_{r_a r_b}(f)$,

$$\frac{G_{r_a r_b}(f)}{|G_{r_a r_b}(f)|}=e^{j2\pi f\tau} \tag{2.64}$$

and

$$\bar{p}_{\bar{r}_a \bar{r}_b}(\zeta) = \delta(\zeta - \tau), \tag{2.65}$$

which is independent of the form of the signal autocorrelation function. Note, however, that when perfect estimation of $G_{r_a r_b}(f)$ is not available, the PHAT GC estimation peak will still be broadened by an amount proportional to the deviation from the ideal cross-spectral density.

2.4.6 Conventional PN-Correlation Method

Recall that the variance lower bounds in Equation (2.49) show that large BT products are required for better performance. Large BT products are desirable when estimating the delay TOAs of a transmitted signal in a multipath channel in order to cover a large interval of frequencies over a large observation window. The large bandwidth is desired so that all frequency components of the channel are perturbed. Large bandwidth usually implies the ability to resolve multipath arrivals separated by $|\tau_2 - \tau_1| \geq \frac{1}{B}$. The large duration is desired because longer signals usually yield higher processing gains. However, transmitting a long duration signal implies that the observation window time T and the number of samples M will be large and will yield large covariance matrix dimensions together with larger computational complexity. This limitation can be solved using the autocorrelation properties of a PN-sequence. A PN-sequence is a long duration broadband signal with a very large BT product but with a short duration (peaky) autocorrelation function [28]. Figure 2.4 shows the *maximal-length shift register* ($m = 7$) PN-sequence autocorrelation function. The chip rate of this sequence was set to 10 MHz. It is observed that the duration of the main autocorrelation function peak is about two times the chip interval T_c, which is the inverse of the chip rate, in this case $T_c = 100$ ns.

Figure 2.5 shows a conventional time-delay estimation method that involves the transmission of a wide-band PN-sequence $c(t)$ and subsequent correlation of the sampled, filtered, and received signal with this sequence [9, 12, 31, 32]. A subset of the local extrema of the resultant magnitude of the correlator output is selected as the multipath delays, and the first delay or the strongest delay may be chosen as the first TOA estimate. Due to the triangular spread of the PN-sequence autocorrelation function, arrivals separated by less than T_c seconds will not be resolved. Thus, the resolution of this method is limited by the chip rate of the signal $c(t)$.

Figure 2.6 shows the power delay profiles, defined as $|z(\zeta)|^2$, of two different two-path channels ($D = 2$). The first channel has delays at $D = [25\,225]$ ns,

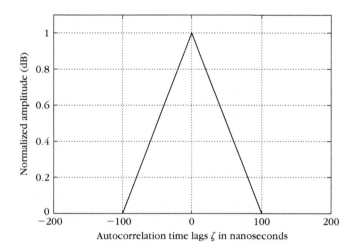

FIGURE 2.4

An $m = 7$ maximal PN-sequence autocorrelation function. $T_C = 100$ ns.

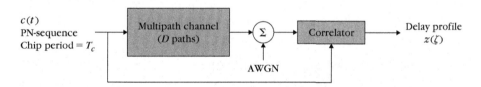

FIGURE 2.5

Conventional measurement of a baseband, multipath channel-delay profile.

a separation of $2T_c$. The second channel has delays at $D = [25\,75]$ ns, a separation of $0.5\,T_c$. For this example, $T_c = 100$ ns. Clearly, the conventional PN-correlation method fails to resolve the delays of the second channel. In this scenario, the method yields a first TOA estimate of 50 ns (first peak of the correlation function), which is 25 ns away from the true value. An obvious and easy solution to this limited resolution problem is to increase the chip rate of $c(t)$ to reduce the base width of the autocorrelation function; however, this is not always feasible since we may have bandwidth limitations or hardware complexity constraints.

Another problem faced by the conventional delay profile measurement is that of mutual path cancellations due to differences between their relative phases. These cancellations cause the peaks of the delay profile to move or to completely disappear from the true delay locations.

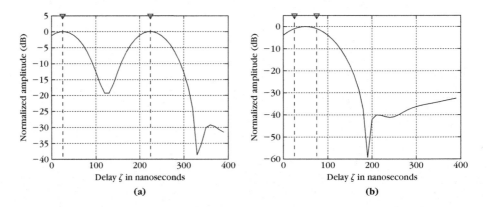

FIGURE 2.6

Power delay profiles of two different two-path channels. The first channel has delays at $D = [25\ 225]$ ns, a separation of $2\ T_c$. The second channel has delays at $D = [25\ 75]$ ns, a separation of $0.5\ T_c$. For this example, $T_c = 100$ ns.

A solution that allows us to keep the signal chip rate untouched and that avoids the path cancellations due to their relative phase difference consists of applying a super-resolution method to the received delay profile $z(\zeta)$.

2.4.7 A Super-Resolution PN-Correlation Method: The SPM Algorithm

This method was presented in [24, 26], and its performance under real tranmission conditions was analyzed in [3]. Consider transmitting a signal through a multipath channel consisting of D discrete paths with various time delays $\{\tau_i; i = 1, \ldots, D\}$ that we want to estimate, and with an impulse response as given in Equation (2.42). If the baseband modulation signal is a PN-sequence $c(t)$ and ω_c is the carrier frequency, then the baseband received signal is given by

$$r(t) = \sum_{i=1}^{D} m_i c(t - \tau_i) + \eta(t), \tag{2.66}$$

where it is assumed that $\eta(t)$ is a stationary, zero mean, additive white-noise process with variance equal to σ^2. The cross-correlation of $r(t)$ with $c(t)$ is

$$z(\zeta) = \frac{1}{T} \int_0^T c(t - \zeta) r(t)^* dt, \tag{2.67}$$

where T was previously described as the snapshot measurement window that starts at the beginning of each PN-sequence transmission and ends when no more echoes arrive at the receiver. Substituting Equation (2.66) into Equation (2.67) yields

$$z(\zeta) = \sum_{i=1}^{D} m_i p(\zeta - \tau_i) + v(\zeta), \qquad (2.68)$$

where

$$p(\zeta) = \frac{1}{T} \int_0^T c(t - \zeta) c(t)^* dt, \qquad (2.69)$$

$$v(\zeta) = \frac{1}{T} \int_0^T c(t - \zeta) \eta(t)^* dt. \qquad (2.70)$$

Thus, we have a new set of functions (autocorrelation and cross-correlation functions) that are defined in the lag domain ζ. It is clear that the delay information we are looking for is inherent in Equation (2.68). So it is very reasonable to consider the signal at the output of the correlator (i.e., the delay profile $z(\zeta)$ signal) as our new received signal and apply the MUSIC algorithm to this new type of data.

Let us sample the delay profile signal at M lags $\{\zeta_1, \zeta_2, \ldots, \zeta_M\}$ to obtain a delay profile vector $\mathbf{z} = [z(\zeta_1)\, z(\zeta_2) \ldots z(\zeta_M)]^T$. It is assumed that the sampling interval is chosen to be a small fraction of the chip interval T_c. From Equation (2.68),

$$\mathbf{z} = \sum_{i=1}^{D} m_i \mathbf{p}(\tau_i) + \mathbf{v}, \qquad (2.71)$$

where our new "steering vector," $\mathbf{p}(\tau)$, is defined as

$$\mathbf{p}(\tau) = [p(\zeta_1 - \tau)\, p(\zeta_2 - \tau) \ldots p(\zeta_M - \tau)]^T. \qquad (2.72)$$

This steering vector is easy to calculate for any τ since the autocorrelation function $p(\zeta)$ is known. The new noise vector is given by $\mathbf{v} = [v(\zeta_1)\, v(\zeta_2) \ldots v(\zeta_M)]^T$. If we define the array manifold matrix and the path gain vector, respectively, as

$$\boldsymbol{\Gamma} = [\mathbf{p}(\tau_1)\, \mathbf{p}(\tau_2) \ldots \mathbf{p}(\tau_D)], \qquad (2.73)$$

and

$$\mathbf{g} = [m_1\, m_2 \ldots m_D]^T, \qquad (2.74)$$

then we can write the delay profile vector in matrix notation as

$$\mathbf{z} = \boldsymbol{\Gamma}\mathbf{g} + \mathbf{v}. \qquad (2.75)$$

Finally, the correlation matrix of this data vector $\mathbf{z}$ is expressed as

$$\mathbf{R} = E\{\mathbf{z}\mathbf{z}^H\} = \sum_{i,j} E\{g_i g_j^*\}\mathbf{p}(\tau_i)\mathbf{p}(\tau_j)^H + E\{\mathbf{v}\mathbf{v}^H\}. \qquad (2.76)$$

In matrix notation,

$$R = \Gamma G \Gamma^H + R_n, \qquad (2.77)$$

where G is the path gain covariance matrix. The noise correlation matrix can be expressed as

$$E\{\boldsymbol{\nu}\boldsymbol{\nu}^H\} = R_n = \sigma^2 R_0. \qquad (2.78)$$

Note that

$$E\{\nu(\zeta_k)\nu(\zeta_l)^*\} = \left\{ \frac{1}{T}\int_0^T E\{\eta(t)\eta(t)^*\}dt \right\} \left\{ \frac{1}{T}\int_0^T c(t-\zeta_k)c(t-\zeta_l)^* dt \right\} = \sigma^2 p(\zeta_k - \zeta_l),$$

$$(2.79)$$

so R_0 is a Hermitian matrix whose kl-th element is equal to $p(\zeta_k - \zeta_l)$, and

$$[R_n]_{k,l} = \sigma^2 p(\zeta_k - \zeta_l). \qquad (2.80)$$

From the noise covariance matrix, we observe that now we deal with a colored noise process, so a whitening procedure should be implemented. This may be achieved by applying the linear transformation $\bar{z} = R_0^{-\frac{1}{2}}z$ so that $\bar{R} = E\{\bar{z}\bar{z}^H\} = R_0^{-\frac{1}{2}}RR_0^{-\frac{1}{2}} = [R_0^{-\frac{1}{2}}\Gamma]G[\Gamma^H R_0^{-\frac{1}{2}}] + \sigma^2 I$. Then solving the generalized eigenvalue problem for the whitened observations yields

$$\bar{R}\bar{e}_l = [R_0^{-\frac{1}{2}}RR_0^{-\frac{1}{2}}]\bar{e}_l = \lambda_l \bar{e}_l, \qquad (2.81)$$

which, clearly, leads to the generalized eigenvalue problem of the form

$$Re_i = \lambda_i R_0 e_i. \qquad (2.82)$$

It may be concluded that solving the generalized eigenvalue problem in Equation (2.82) is equivalent to whitening the noise [45].

Again we see that the structure of Equations (2.75) and (2.77) is identical to Equations (2.5) and (2.7), respectively. Table 2.2 shows the analogies of the three estimation cases we have discussed in this section and the previous one.

We should make it clear that to apply MUSIC to this new model, the steering vectors $\{\mathbf{p}(\tau_1); \dots; \mathbf{p}(\tau_D)\}$ should form a linearly independent set so that Γ is full rank. Since the assumption of linear independence between steering vectors is vital for SPM, let us illustrate the degree of dependence now by creating a two-path channel with delays at τ_1 and $\tau_1 + \Delta\tau$ seconds. Using a 10-MHz, chip-rate m-sequence ($T_c = 100$ ns), we calculate a set of array manifold matrices Γ and plot the condition number of $\Gamma^H\Gamma$ as a function of $\Delta\tau$. Figure 2.7 shows this plot. As expected, for $\Delta\tau = 0$ the smallest eigenvalue is zero since the rank of Γ becomes one. However, for values of $\Delta\tau > 0$, the smallest eigenvalue has

Table 2.2 Analogies between Methods

Method	AOA	Time Delay	SPM
Signal	$r_m(t) = \sum_{i=1}^{D} s_i(t)e^{-j(m-1)k_i} + \eta$	$r(t) = \sum_{i=1}^{D} m_i s(t - \tau_i) + \eta$	$z(\zeta) = \sum_{i=1}^{D} m_i p(\zeta - \tau_i) + \nu$
M	Number of sensors	Number of time samples	Number of lag samples
D	Number of impinging waves	Number of channel paths	Number of channel paths
Steering vector	$a_k = [1\,e^{jk}\,\ldots\,e^{(M-1)jk}]^T$	$a(\tau) = [s(t_1 - \tau)\,\ldots\,s(t_M - \tau)]^T$	$p(\tau) = [p(\zeta_1 - \tau)\,\ldots\,p(\zeta_M - \tau)]^T$
Model	$r(t) = As(t) + \eta(t)$	$r = Ag + \eta$	$z = \Gamma g + \nu$
Correlation matrix	$ASA^H + R_n$	$AGA^H + R_n$	$\Gamma G \Gamma^H + R_n$
Noise-correlation matrix	$R_n = \sigma^2 I$	$R_n = \sigma^2 I$	$R_n = \sigma^2 R_0$
Parameter of interest	$\{\theta_i;\ i = 1,\ldots,D\}$	$\{\tau_i;\ i = 1,\ldots,D\}$	$\{\tau_i;\ i = 1,\ldots,D\}$

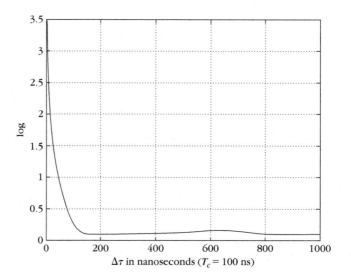

FIGURE 2.7

Condition number of the matrix $\boldsymbol{\Gamma}^H \boldsymbol{\Gamma}$ as a function of the delay difference $\Delta \tau$ seconds of a two-path channel.

non-zero values that grow very fast with respect to $\Delta \tau$. This indicates that, at least in certain conditions, $\boldsymbol{\Gamma}$ is a full-rank matrix when the delay difference is a small fraction of the chip period [3].

In all our discussions thus far, once the correlation matrix was obtained, we could directly apply MUSIC since we assumed that the multipath return signals were uncorrelated—that is, that the path-gain correlation matrix $\mathbf{G}$ was a diagonal matrix. This is usually not true, especially in slow time-varying channels where the signals will be correlated since the channel path magnitudes and relative phases will remain constant during the acquisition of the N snapshots. This leads to a rank-deficient matrix $\mathbf{R} - \mathbf{R_n}$. To solve this problem, a decorrelating technique similar to spatial smoothing, called *frequency smoothing*, may be used to restore the rank of $\mathbf{R} - \mathbf{R_n}$. This technique consists of transmitting the PN-sequence using distinct carrier frequencies [3, 24].

Assuming decorrelated multipath arrivals, MUSIC may be applied as follows to obtain the SPM algorithm [24]:

1. Estimate the signal correlation matrix $\mathbf{R}$ using N snapshots.

2. Find the generalized eigenvalues and eigenvectors of $\hat{\mathbf{R}}$.

3. Find the noise subspace spanned by $\{\mathbf{e}_i; i = D + 1, D + 2, \ldots, M\}$, which are the generalized eigenvectors corresponding to the $M - D$ smallest eigenvalues.

4. Calculate the MUSIC spectrum, that is, the *super-resolution delay profile* (SDP):

$$\text{SDP}(\tau) = \frac{\mathbf{p}(\tau)^H \mathbf{R}_0^{-1} \mathbf{p}(\tau)}{\sum_{i=D+1}^{M} |\mathbf{p}(\tau)^H \mathbf{e_i}|^2}. \tag{2.83}$$

The SDP will ideally peak to infinity when the true delay values $\{\tau_i;\ i = 1, \ldots, D\}$ of the channel are input into this function. The first TOA estimate may be chosen as the first peak in the SDP.

In an analogy to the array calibration errors in the MUSIC-based AOA estimation problem, MUSIC-based TOA estimation will suffer from unknown filter transfer functions at the transmitter and receiver, which may distort the form of the correlation function and change the form of the "steering vectors" [24]. A way to overcome this problem is to include the filtering distortions in the calculation of the steering vectors $\mathbf{p}(\tau)$. If $h_F(t)$ is the cascade impulse response of all the filtering operations on the signal prior to the SPM processing, then we can modify Equation (2.69) as

$$\tilde{p}(\zeta) = \frac{1}{T} \int_0^T c(t - \zeta)[h_F(t) \otimes c(t)]dt = h_F(\zeta) \otimes p(\zeta), \tag{2.84}$$

and calculate the steering vector in Equation (2.72) using $\tilde{p}(\zeta)$ instead of $p(\zeta)$.

The SDP of four different uncorrelated two-path channels with $|\tau_2 - \tau_1| = \{0.25\,T_c, 0.2\,T_c, 0.1\,T_c, 0.05\,T_c\}$ were obtained at different SNR scenarios. The paths were created with equal amplitudes and zero degrees of relative phase between them, and 10-MHz PN-sequences were used for transmission, which means that the resolution of the conventional correlation method will be limited to 100 ns. The sampling interval at the output of the correlator was set to $0.1T_c$. Note that we will try to measure a delay separation of $0.05\,T_c$, which is smaller than this sampling interval. The noise covariance matrix was calculated theoretically using Equation (2.80), and $N = 150$ snapshots were used to estimate the signal covariance matrix. Figures 2.8, 2.9, and 2.10 show the results for SNRs of 40 dB, 30 dB, and 10 dB, respectively, where SNRs are measured at the correlator output. The true locations of the path delays are marked with arrows. The solid plots correspond to the super-resolution delay profiles of the channels, and the dotted plots correspond to the channel power delay profiles $|z(\zeta)|^2$ obtained with the conventional PN-correlation method.

From these plots it can be seen that for high SNRs we can correctly estimate the true delay locations of the channels, even for path separations smaller than the sampling interval at the output of the correlator. For low SNRs we can still resolve paths separated by $0.2\,T_c$ seconds or more but, because of the large noise, now the estimates have errors and the SDPs do not peak at the exact true

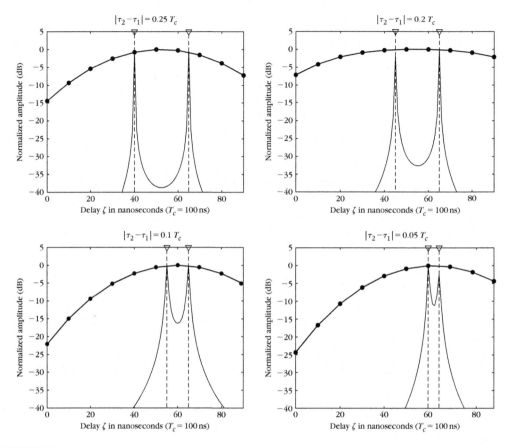

FIGURE 2.8

Super-resolution delay profiles for two-path channels with **SNR** = 40 dB measured at the correlator output. The dotted plots are the power delay profiles measured at the correlator output; the true path delay locations are marked by arrows.

delay locations. Additive noise lowers the resolution capabilities and causes the SDP estimates to have bias errors. Note how the dotted-line plots corresponding to the conventional correlation method are far from resolving two paths in any of the examples.

The PN-correlation–based methods presented in this section and the previous one choose the first TOA as the first peak on the estimation function (either the correlator output in the conventional case or the SDP in its super-resolution counterpart). These methods commonly suffer from two major sources of error. The first is LOS attenuation, which may be caused by shadowing. When this happens, a contiguous, subsequent strongest non-LOS arrival may be chosen as the first TOA, and this in turn will translate into range estimation errors. The

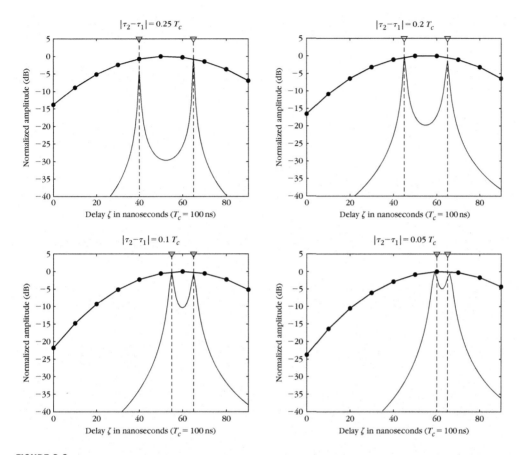

FIGURE 2.9

Super-resolution delay profiles for two-path channels with **SNR** = 30 dB measured at the correlator output. The dotted plots are the power delay profiles measured at the correlator output; the true path delay locations are marked by arrows.

second one is limited resolution, which may cause multipaths that arrive less than T_c seconds after the LOS arrival time to add up destructively and attenuate the magnitude of the LOS path. Unresolvable multipath components usually shift the estimator peaks to values away from the true first TOA, causing estimation errors.

TOA-based systems require high clock synchronization between all the system components in order for a receiver to know the exact time when a signal was transmitted by a source. Poor synchronization (clock drifts) is another major source of error in TOA-based systems.

TDOA systems avoid the requirement of clock synchronization at the point of interest or tag-end by considering differences in TOA of signals that originate at

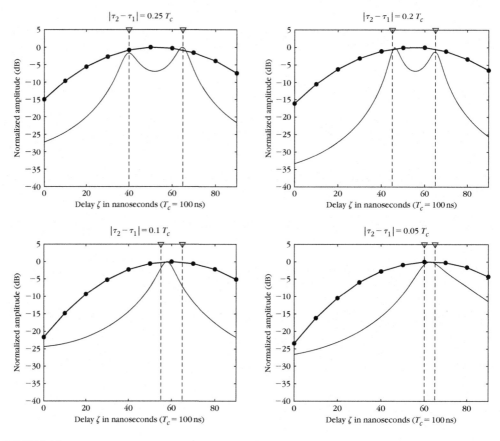

FIGURE 2.10

Super-resolution delay profiles for two-path channels with **SNR = 10 dB** measured at the correlator output. The dotted plots are the power delay profiles measured at the correlator output; the true path delay locations are marked by arrows.

two different reference points. The TDOA calculation will effectively cancel time synchronization errors at the tag. Note that TDOA-based positioning schemes still require clock synchronization among all the receivers in the system.

2.4.8 TOA Estimation by Successive Cancellation

A whitening operation was discussed in the previous section, which led to substitution of the simple eigenvalue problem to a generalized one. The whitening idea has been used effectively to improve SPM under noisy scenarios. For instance, Bouchereau et al. [3] discusses the performance of MUSIC in the presence of narrow-band jammers. In this work, it was concluded that if an estimate of the covariance matrix of the ensemble of additive and jammer noise as seen at the output of the correlator was available, then it could be included in the generalized eigenvalue problem (as described in Equation (2.82)) to effectively eliminate the jamming effects. This could be stated as

$$\mathbf{Re_i} = \lambda_i[\mathbf{R_0} + \mathbf{R_J}]\mathbf{e}_i, \qquad (2.85)$$

where $\mathbf{R_0}$ has already been defined as the normalized correlator output noise covariance matrix, and $\mathbf{R_J}$ is a covariance matrix of any other noise process present in the channel (for the jamming example, $\mathbf{R_J}$ would stand for the jamming process covariance matrix). This same idea was used in Krasny and Koorapaty [20] to obtain an improved MUSIC and conventional correlation estimator for TOA; the technique was named *successive cancellation* by the authors.

Successive cancellation is an iterative algorithm. At the i-th iteration, a channel delay profile is estimated by MUSIC (or SPM if a PN-sequence correlation is used as the received signal) or by the conventional correlation method. The position of the D largest peaks of this profile are then chosen as delay estimates $\{\hat{\tau}_1^i, \ldots, \hat{\tau}_D^i\}$, where the superscript has been used to denote the iteration number. Then at the next iteration a new noise correlation matrix is formed as

$$\mathbf{R_n}^{(i+1)} = \mathbf{R_n}^{(i)} + \hat{\mathbf{A}}^{(i)}\mathbf{G}\hat{\mathbf{A}}^{(i)H}, \qquad (2.86)$$

where $\hat{\mathbf{A}}^{(i)}\mathbf{G}\hat{\mathbf{A}}^{(i)H}$ corresponds to the signal correlation term as given in Equation (2.54) which is evaluated at the delays estimated at the i-th iteration such that

$$\hat{\mathbf{A}}^{(i)} = [\mathbf{a}(\hat{\tau}_1^{(i)})\,\mathbf{a}(\hat{\tau}_2^{(i)}) \cdots \mathbf{a}(\hat{\tau}_D^{(i)})], \qquad (2.87)$$

where $\mathbf{a}(\tau)$ is a steering vector as defined in Equation (2.53). As it can be seen, at each iteration, the algorithm treats delayed transmitted signals $\{s(t - \hat{\tau}_j^{(i)}), i = 1, \ldots, D\}$ as noise and attempts to cancel them out by embedding their correlation matrix in the overall noise matrix. This is equivalent to attempting to eliminate multipath components that may be overlapping in the delay profile function. For the case of MUSIC, this cancellation is achieved by solving, at each iteration, a generalized eigenvalue problem of the form

$$\mathbf{Re}^{(i+1)} = \lambda^{(i+1)}\mathbf{R_n}^{(i+1)}\mathbf{e}^{(i+1)}. \qquad (2.88)$$

The delay profile of the channel and the corresponding D delay estimates may then be estimated with the resulting generalized eigenvectors using Equation (2.25).

For the case of the conventional correlation delay estimation method, the generalized correlation function given by

$$p^{(i+1)}(\tau) = \frac{1}{N}\sum_{i=1}^{N}|[\mathbf{R_n}^{(i+1)}]^{-1}\mathbf{a}(\tau)^H\mathbf{r}_i|^2, \qquad (2.89)$$

may be used to suppress delay components at $\{\hat{\tau}_1^{(i)}, \ldots, \hat{\tau}_D^{(i)}\}$ from the correlation function estimate. This last function uses N received signal snapshots $\{\mathbf{r}_i, i = 1, \ldots, N\}$ as described in Equation (2.52).

In the first iteration, the noise correlation matrix will be simply $\sigma^2\mathbf{I}$ (or $\sigma^2\mathbf{R_0}$ for SPM). After the last iteration, all the time delays for which a signal was detected are tabulated and the earliest delay is chosen as the first TOA of the received signal. In the tabulation process, all detected delays that lie outside a window $[\tau_p - \delta, \tau_p + \delta]$ are ignored. Here, τ_p is the delay at which either the correlation function or MUSIC pseudospectrum presented the maximum peak at the first iteration. The iterations may be stopped when every peak on the delay profile estimate is below a threshold value. It is clear that this algorithm will be strongly influenced by the capacity of the delay profile estimation method to resolve two consecutive paths; this capacity is measured using a probability of resolution metric. An excellent discussion on MUSIC's probability of resolution can be found in Zhang [50].

The successive cancellation algorithm may effectively improve TOA estimation performance in the presence of multipath components, as has been shown in Krasny and Koorapaty [21].

2.5 RANGE ESTIMATION BASED ON RECEIVED SIGNAL STRENGTH

RSS measurements are useful in PL systems since they may be related to distance between a transmitter and a receiver through an adequate propagation model. Although signal strength measurements are quite simple to obtain, obtaining an accurate propagation model might be the opposite. Propagation models are usually based on extensive measurement campaigns and are strongly dependent on the specific scenario (indoors, outdoors, heavy clutter, etc.), frequency band, weather conditions, and sometimes even time of day.

RSS measurements are affected by two major channel impairments known as large-scale fading and small-scale fading (see [32]), which are usually treated as random processes. Ideally, the two fading processes need to be considered when obtaining a propagation model, and this may be done by considering compound fading models [6, 13, 14]. However, due to the mathematical complexity involved in the estimation of the parameters of these schemes, it is common to ignore one of the fading processes or to reduce its impact on the signal strength measurements by means of temporal and spatial averaging as the following explains.

In this section, we will discuss ways to measure received signal strength based on spatial samples or on the channel's power delay profile. We will also discuss averaging techniques to eliminate small-scale fading effects. Finally, we will introduce the log-normal path-loss model that is widely used in RSS-based PL schemes. Here we will present ML estimators to find the log-normal parameters based on a set of power measurements and, subsequently, an estimator of range based on these parameters.

2.6 RECEIVED SIGNAL STRENGTH MEASUREMENTS

The power of a received signal over a flat-fading wireless channel is composed of a fast-fading superimposed on a large-scale faded signal [6, 13, 14, 22]. Then the received signal power may be modeled as

$$p_r(t) = \gamma(t)p_0(t), \tag{2.90}$$

or

$$p_r(l) = m(l)p_0(l), \tag{2.91}$$

where the two equations represent the signal as measured at time t or at position l, respectively (clearly these two quantities may be related through the receiver's velocity). The objective is to find an estimate of $\gamma(t)$, which refers to the local average power, by eliminating the small-scale fading term $p_0(t)$. It is common to assume that the two fading processes are statistically independent and that $E\{p_0\} = 1$ [6, 14] (this last assumption causes no loss of generality in the model since a nonunity small-scale fading gain will be absorbed by the average of the large-scale fading component). Further, it is usually assumed that $\gamma(t)$ follows a log-normal distribution, whereas $p_0(t)$ may be described by a Rayleigh, Ricean, Suzuki, and Nakagami-m distribution, among others [6].

RSS may be estimated by obtaining the spatial average given by

$$\hat{\gamma}(l) = \frac{1}{2L} \int_{l-L}^{l+L} p_r(\alpha)d\alpha. \tag{2.92}$$

Since it is generally accepted that a log-normal fading component is constant over a spatial dimension that exceeds a few hundred wavelengths [43], it is reasonable to assume that in the PL estimation problem $m(l) = m$, Equation (2.92) becomes

$$\hat{\gamma}(l) = \frac{\gamma}{2L} \int_{l-L}^{l+L} p_0(\alpha)d\alpha. \tag{2.93}$$

Clearly, $\hat{\gamma} \to \gamma$ whenever the integral in Equation (2.93) becomes unity, which is congruent with the fact that $E\{p_0\} = 1$. The variance of the above RSS estimator was found in [22] for the case of Rayleigh small-scale fading and is given by

$$\text{VAR}(\hat{\gamma}) = \frac{\lambda}{4L}\sigma_R \sqrt{\frac{\pi}{2}} \int\limits_0^{2L\lambda} \left(1 - \frac{\lambda\alpha}{2L}\right) J_0^2(2\pi\alpha)d\alpha, \tag{2.94}$$

where $\sigma_R = \frac{1}{2}E\{p_r^2\}$, and $J_0(\cdot)$ is the zero-order Bessel function of the first kind. If L is made too short, the estimate will have large variance since it will still contain small-scale fading information. On the other hand, large values of L risk the smoothing out of large-scale fading information. An averaging over 20 to 40 wavelengths is recommended in [22].

Whenever N discrete, uncorrelated RSS measurements are available at locations $\{l_i, i = 1, \ldots, N\}$, the RSS estimator becomes

$$\hat{\gamma}_{dB} = \frac{1}{N}\sum_{i=1}^N p_{r_{dB}}(l_i), \tag{2.95}$$

where subscript dB denotes that the quantities are expressed in dBs. Note then that, by the central limit theorem, for large N, $\hat{\gamma}$ will be log-normal regardless of the distribution of p_r. The mean and variance, respectively, of this estimator are found in [22] as

$$E\{\hat{\gamma}\} = \gamma,$$
$$\text{VAR}(\hat{\gamma}) = O\left(\frac{1}{N}\right). \tag{2.96}$$

According to [22], a value of $N \geqslant 36$ will achieve a $\hat{\gamma}$ that lies within ± 1 dB of its true mean. Finally, the uncorrelated sample assumption will hold whenever the distance between measurements is greater than or equal to $0.8\,\lambda$.

As a final comment, we note that elimination of small-scale fading effects is achieved through spatial averaging, which implies that measurements must be obtained at distinct locations. If this is not possible, then large-scale fading effects will be difficult to eliminate and a range estimator based on the compound fading model in Equation (2.90) may be necessary to increase accuracy. Parameter estimators for compound fading models have been presented in [6, 13, 14, 22], for example.

Now let us consider a multipath channel with D resolvable paths with impulse response $h(t)$ and a received signal $r(t)$, as given in Equations (2.42) and (2.43), respectively. In this scenario, the instantaneous received signal power at the j-th snapshot time t_j may be expressed [32] as

$$p_r(t_j) = \frac{1}{\tau_D} \int\limits_0^{\tau_D} |r(t)|^2 dt = k\sum_{i=1}^D |m_{ij}|^2, \tag{2.97}$$

where k is a constant that depends on the amplitude and duration (energy) of the transmitted signals. Taking the expectation of this equation and normalizing with respect to k yields the average received signal strength for the multipath channel as

$$\bar{\gamma} = \sum_{i=1}^{D} E\{|m_i|^2\} \approx \sum_{i=1}^{D} \gamma_i. \qquad (2.98)$$

Note that if the power delay profile of the channel is obtained at N different uncorrelated locations, γ_i, the local average power of the i-th path may be obtained as explained previously using Equation (2.92) or (2.95). Either equation would have to be applied for each of the D resolvable paths observed on each of the N available power delay profiles.

2.6.1 Log-Normal Propagation Model

The log-normal path-loss model may be considered as a generalization of the free-space Friis equation [32] where the power is allowed to decrease at a rate of $(1/d)^n$ (where d denotes distance or range), and where a random variable is added in order to account for shadowing (large-scale fading) effects. The model may be expressed as

$$p_r(d)_{dB} = \bar{p}_r(d_0)_{dB} - 10n \, \log\left(\frac{d}{d_0}\right) + \chi, \quad d > d_0, \qquad (2.99)$$

where log stands for base 10 logarithm, $p_r(d)_{dB}$ is the received power at a distance d meters from the transmitter, n is the path-loss exponent that defines the rate of decay of power with respect to distance, and χ is a Gaussian random variable with zero mean and variance σ_χ^2 that is defined in dBs. Further, $\bar{p}_r(d_0)_{dB}$ denotes the ensemble average over all possible received power values for a given reference distance denoted as d_0 meters. This reference power is usually measured a priori or calculated with the free space Friis equation. In all cases, d_0 should be as small as possible while being in the far-field region from the transmitter. In our discussion, we will consider that power measurements exclude small-scale fading since they were obtained by the averaging techniques presented in Section 2.6. The log-normal model basically states that the received power is not uniform when measured at different locations while maintaining the same distance separation between the transmitter and receiver. Note that a Gaussian random variable defined in decibels becomes a log-normal random variable when transformed into the linear domain. The log-normal Equation (2.99) is clearly a line with slope $10n$ when plotted versus distance values given in decibels. Typical values of path-loss exponents range between 1.5 and 5.

2.6.2 ML Estimation of Log-Normal Parameters

Parameters n and σ_χ^2 are usually unknown and must be estimated based on channel measurements. Once these parameters are obtained, distance estimation from the log-normal model becomes relatively straightforward. Let us then obtain expressions for the estimates of n and σ_χ^2.

Consider that a set of N average power observations have been obtained at different distances from a transmitter and at different uncorrelated locations in the area of interest such that N measurement pairs $\{p_r(d_i), d_i\}$ are available. For simplicity we will drop the subscript dB and assume that all power measurements are given in decibels. The log-likelihood function is obtained from the joint probability density of all N power observations as

$$L(n, \sigma_\chi^2) = -N\ln\left(\sqrt{2\pi\sigma_\chi^2}\right) - \frac{1}{2\sigma_\chi^2} \sum_{i=1}^{N} \left[p_r(d_i) - p_r(d_0) - 10n \log\left(\frac{d_i}{d_0}\right)\right]^2. \quad (2.100)$$

Differentiating with respect to n and equating to zero leads to the estimate

$$\hat{n} = \frac{\sum_{i=1}^{N} \log\left(\frac{d_i}{d_0}\right)\left[p_r(d_i) - p_r(d_0)\right]}{10 \sum_{i=1}^{N} \left(\log\left(\frac{d_i}{d_0}\right)\right)^2}. \quad (2.101)$$

Substituting n by its estimate $\hat{n}$ in Equation (2.100), taking derivatives with respect to σ^2 and equating to zero, we obtain

$$\hat{\sigma}_\chi^2 = \frac{1}{N} \sum_{i=1}^{N} \left[p_r(d_i) - p_r(d_0) - 10\hat{n} \log\left(\frac{d_i}{d_0}\right)\right]^2. \quad (2.102)$$

2.6.3 Log-Normal Range Estimator

When the log-normal parameters have been estimated, power measurements may be used to infer the distance between the transmitter and receiver. Substituting estimates $\hat{n}$ and $\hat{\sigma}_\chi^2$ and rewriting Equation (2.100) for a single power observation p_r, we obtain

$$L(\hat{m}, \hat{\sigma}_\chi^2, d) = -\ln\left(\sqrt{2\pi\hat{\sigma}_\chi^2}\right) - \frac{1}{2\hat{\sigma}_\chi^2} \left[p_r - p_r(d_0) - 10\hat{n} \log\left(\frac{d}{d_0}\right)\right]^2. \quad (2.103)$$

Taking a derivative with respect to d and equating to zero yields the distance estimator

$$\hat{d} = d_0 10^{\left[\frac{p_r(d_0) - p_r}{10\hat{n}}\right]}. \quad (2.104)$$

Let us now calculate the expected value and variance of this distance estimator assuming that $\hat{n}$ is a constant (i.e., not random). First, we note from Equation (2.99) that $p_r(d_0) - p_r = 10\hat{n} \log\left(\frac{d}{d_0}\right) + \chi$, and thus the estimator in (2.104) may be rewritten as

$$\hat{d} = d \, 10^{\frac{\chi}{10\hat{n}}}. \tag{2.105}$$

Note that if $\chi \sim N(0, \sigma_\chi^2)$ and $10 \log \xi \sim N(0, \sigma_\chi^2)$, then $\xi = e^{\frac{\chi}{10 \log e}}$ is a log-normal distributed random variable. Equation (2.105) may then be redefined in terms of this variable as

$$\hat{d} = d\xi^{1/\hat{n}} = d(e^{\frac{\chi}{10 \log e}})^{1/\hat{n}}. \tag{2.106}$$

Finally, it follows from the Gaussian moment-generating function [27] that

$$E\{\hat{d}\} = d \exp\left[\frac{\sigma_\chi^2}{200\hat{n}^2 \log^2 e}\right], \tag{2.107}$$

$$\text{VAR}\{\hat{d}\} = d^2 \left[\exp\left(\frac{\sigma_\chi^2}{50\hat{n}^2 \log^2 e}\right) - \exp\left(\frac{\sigma_\chi^2}{100\hat{n}^2 \log^2 e}\right)\right], \tag{2.108}$$

which show that the estimator is biased. An unbiased estimator is readily obtained by letting

$$\tilde{d} = \frac{\hat{d}}{c}, \tag{2.109}$$

where

$$c = \exp\left[\left(\frac{\sigma_\chi^2}{200 n^2 \log^2 e}\right)\right], \tag{2.110}$$

which yields a variance of

$$\text{VAR}\{\tilde{d}\} = d^2 \left[\exp\left(\frac{\sigma_\chi^2}{100\hat{n}^2 \log^2 e}\right) - 1\right]. \tag{2.111}$$

Note that the variance of the estimates of range increases with range d. This property is common to RSS-based range estimation schemes.

Distance estimation based on RSS measurements will suffer from non–line-of-sight scenarios (shadowing), where the received power will be reduced by attenuation due to wave propagation through an obstacle.

REFERENCES

[1] M. Azaria, D. Hertz, Time delay estimation by generalized cross correlation methods, IEEE Transaction on Acoustics, Speech, and Signal Processing, 32 (2) (1984) 280–285.

[2] M. Bekara, M. Van Der Baan, A new parametric method for time delay estimation, IEEE International Conference on Acoustic Speech and Signal Processing, 3 (2007) 1033–1036.

[3] F. Bouchereau, D. Brady, C. Lanzl, Multipath delay estimation using a superresolution PN-correlation method, IEEE Transactions on Signal Processing, 49 (5) (2001) 938–949.

[4] F. Bouchereau, D. Brady, Bounds on range-resolution degradation using RSSI measurements, in: Proceedings IEEE International Conference on Communications, June (2004) 3246–3250.

[5] F. Bouchereau, D. Brady, Resolution bounds for statistical location estimation in a hallway, in: Proceedings of the International Workshop on Wireless Communications in Underground and Confined Areas, June 2005.

[6] F. Bouchereau, D. Brady, Method-of-moments parameter estimation for compound fading processes, IEEE Transactions on Communications, 56 (2) (2008) 166–172.

[7] A.M. Bruckstein, T.-J. Shan, T. Kailath, The resolution of overlapping echos, IEEE Transactions on Acoustics, Speech, and Signal Processing, 33 (6) (1985) 1357–1367.

[8] X-L. Xu, K.M. Buckley, Bias analysis of the MUSIC location estimator, IEEE Transactions on Signal Processing, 40 (10) (1992) 2559–2569.

[9] R.J.C. Bultitude, S.A. Mahmoud, W.A. Sullivan, A comparison of indoor radio propagation characteristics at 910 MHz and 1.75 GHz, IEEE Journal on Selected Areas in Communications, 7 (1989) 20–30.

[10] G.C. Carter, A.H. Nutall, P.G. Cable, The smoothed coherence transform, IEEE Proceedings, 61 (1973) 1497–1498.

[11] Y. Chengyou, X. Shanjia, W. Dongjin, On the asymptotic analysis of Cramer-Rao bound for time delay estimation, Proceedings of ICSP (1998) 109–112.

[12] D.M.J. Devasirvatham, Time delay and signal measurements of 850 MHz radio waves in building environments, IEEE Transactions on Antennas and Propagation, AP-34 (1986) 1300–1305.

[13] A. Dogandzic, J. Jin, Maximum likelihood estimation of statistical properties of composite gamma-lognormal fading channels, IEEE Transactions on Signal Processing, 52 (10) (2004) 2940–2945.

[14] A. Dogandzic, B. Zhang, Dynamic shadow-power estimation for wireless communications, IEEE Transactions on Signal Processing, 53 (8) (2005) 2942–2948.

[15] B. Frielander, A sensitivity analysis of the MUSIC algorithm, IEEE Transactions on Acoustics, Speech, and Signal Processing, 38 (10) (1990) 1740-1751.

[16] F. Gross, Smart Antennas for Wireless Communications, McGraw-Hill, 2005.

[17] J. Hassab, R.E. Boucher, Optimum estimation of time delay by a generalized correlator, IEEE Transactions on Acoustics, Speech, and Signal Processing, 27 (4) (1979) 373-380.

[18] S. Haykin, Adaptive Filter Theory, Prentice Hall, 1995.

[19] C. Knapp, C. Carter, The generalized correlation method for estimation of time delay, IEEE Transactions on Acoustics, Speech, and Signal Processing, 24 (4) (1976) 320-327.

[20] L. Krasny, H. Koorapaty, Enhanced time of arrival estimation with successive cancellation, IEEE Vehicular Technology Conference, 6 (2002) 851-855.

[21] L. Krasny, H. Koorapaty, Performance of successive cancellation techniques for time of arrival estimation, IEEE Vehicular Technology Conference, 4 (2002) 2278-2282.

[22] W.C.Y. Lee, Estimate of local average power of a mobile radio signal, IEEE Transactions on Vehicular Technology, 34 (1) (1985).

[23] X. Li, K. Pahlavan, J. Beneat, Performance of TOA estimation techniques in indoor multipath channels, IEEE International Symposium on Personal Indoor and Mobile Radio Communications, 2 (2002) 911-915.

[24] T. Manabe, H. Takai, Superresolution of multipath delay profiles measured by PN correlation method, IEEE Transactions on Antennas and Propagation, 40 (5) (1992) 500-509.

[25] B. Ottersten, M. Viberg, T. Kailath, Performance analysis of the total least squares ESPRIT algorithm, IEEE Transactions on Signal Processing, 39 (5) (1991) 1122-1134.

[26] M-A. Pallas, G. Jourdain, Active high resolution time delay estimation for large BT signals, IEEE Transactions on Signal Processing, 39 (4) (1991) 781-787.

[27] A. Papoulis, S.U. Pillai, Probability, Random Variables and Stochastic Processes, McGraw-Hill, 2001.

[28] R.L. Peterson, R.E. Ziemer, D.E. Borth, Introduction to Spread Spectrum Communications, Prentice Hall, 1995.

[29] J.G. Proakis, D.G. Manolakis, Digital Signal Processing, Principles, Algorithms, and Applications, Prentice Hall, 2006.

[30] A. Quazi, An overview on time delay estimation in active and passive systems for target localization, IEEE Transactions on Acoustics, Speech, and Signal Processing, 29 (3) (1981) 527-533.

[31] T.S. Rappaport, Characterization of UHF multipath radio channels in factory buildings, IEEE Transactions on Antennas and Propagation, AP-37 (1989) 1058-1069.

[32] T.S. Rappaport, Wireless Communications, Principles and Practice, Prentice Hall, 2002.

[33] V.U. Reddy, A. Paulraj, T. Kailath, Performance analysis of the optimum beamformer in the presence of correlated sources and its behavior under spatial smoothing, IEEE Transactions on Acoustics, Speech, and Signal Processing, ASSP-35 (7) (1987) 927–935.

[34] K. Maheswara Reddy, V.U. Reddy, Further results in spatial smoothing, Signal Processing, 18 (1965) 217–224.

[35] P.R. Roth, Effective measurements using digital signal analysis, IEEE Spectrum, 8 (1971) 62–70.

[36] R. Roy, T. Kailath, ESPRIT- estimation of signal parameters via rotational invariance techniques, IEEE Transactions on Acoustics, Speech, and Signal Processing, 37 (7) (1989) 984–994.

[37] T-J. Shan, M. Wax, T. Kailath, On spatial smoothing for direction of arrival estimation of coherent signals, IEEE Transactions on Acoustics, Speech, and Signal Processing, ASSP-33 (4) (1985) 806–811.

[38] L.L Scharf, Statistical Signal Processing—Detection, Estimation, and Time Series Analysis, 2nd edition, Addison-Wesley, 1991.

[39] R.O. Schmidt, Multiple emitter location and signal parameter estimation, IEEE Transactions on Antennas and Propagation, AP-34 (1986) 276–280.

[40] P. Stoica, A. Nehorai, MUSIC, maximum likelihood, and Cramer-Rao bound, IEEE Transactions on Acoustics, Speech, and Signal Processing, 37 (5) (1989) 720–741.

[41] P. Stoica, K.C. Sharman, Maximum likelihood methods for direction-of-arrival estimation, IEEE Transactions on Acoustics, Speech, and Signal Processing, 38 (7) (1990) 1132–1142.

[42] P. Stoica, R. Moses, Spectral Analysis of Signals, Pearson-Prentice Hall, 2005.

[43] H. Suzuki, A statistical model for urban radio propagation, IEEE Transactions on Communications, 25 (7) (1977) 673–680.

[44] A. Swindlehurst, T. Kailath, A performance analysis of subspace-based methods in the presence of model errors: part I—The MUSIC algorithm, IEEE Transactions on Signal Processing, 40 (7) (1992) 1758–1774.

[45] C.W. Therrien, Discrete Random Signals and Statistical Signal Processing, Prentice Hall, 1992.

[46] H.L. Van Trees, Optimum Array Processing: Part IV of Detection, Estimation, and Modulation Theory, John Wiley and Sons, 2002.

[47] M. Wax, T. Kailath, Detection of signals by information theoretic criteria, IEEE Transactions on Acoustics, Speech, and Signal Processing, ASSP-33 (2) (1985) 387–392.

[48] M. Wax, I. Ziskind, Detection of the number of coherent signals by the MDL principle, IEEE Transactions on Acoustics, Speech, and Signal Processing, 37 (8) (1989) 1190–1196.

[49] W. Xu, M. Kaveh, Comparative study of the biases of MUSIC-like estimators, Signal Processing, 50 (1996) 39–55.

[50] Q.T. Zhang, Probability of resolution of the MUSIC algorithm, IEEE Transactions on Signal Processing, 43 (4) (1995) 978–987.

Location Information Processing

3

A position location (PL) algorithm provides an estimate of the location or position of a user within a specific frame of reference. It carries out the task of gathering subscriber information in a given coverage area so that a location is estimated by processing this information. The system usually gets information from measurements of the signal transmitted by the node to be located and this information is processed to determine parameters such as the time of arrival (TOA), direction of arrival (DOA), or the received signal strength (RSS), among others, as described in Chapter 2. The location estimate is generally obtained from mappings or geometric relationships that relate such parameters to coordinates, or from the optimal solution to nonlinear and overdetermined systems of equations.

Reconfigurable networks, such as ad hoc and sensor networks, must be aware of these services and applications as well, but their specific characteristics change the traditional PL point of view in a wireless scenario. In these networks, satellite-based positioning, such as GPS, is not a good solution since line-of-sight problems would make indoor positioning practically impossible. In addition to that, the power consumption of satellite-based positioning would reduce node battery life and the use of satellite positioning devices would dramatically increase node cost and size.

Although position location information (PLI) is often provided through GPS, trilateration processes based on range, differential distance, or angular views from known land-fixed reference points present several drawbacks for ad hoc and sensor network environments, becoming more complex as such networks are self-creating, self-organizing, and self-managing [9]. Further, the nodes in these networks are usually power limited such that they may only be able to communicate with their closest neighbors, which may not necessarily correspond to land references.

In many scenarios of interest, such as ad hoc networks, a node of interest (NOI) whose location must be estimated may not have direct connectivity to any or some land references in the network and the direct inference of range or angle of arrival (AOA) from signal measurements may not be possible. In this type of environment, connectivity to a specific NOI may be achieved by the concatenation of multiple links between intermediate devices whose location is also uncertain. In this chapter, we will explore techniques to infer range and/or angle of arrival from previously described multihop links and describe ways to combine the resulting measurements to obtain position estimates of a specific NOI in the network environment.

3.1 THE MULTILATERATION PROBLEM

As previously described, the PL problem involves the observation of range or AOA measurements that are obtained by an NOI via signal transmissions to or from reference nodes or land references. Multiple measurements may be obtained from various land references and subsequently combined in an optimal or suboptimal way to obtain an estimate of the position of the NOI. This process of combining the multiple range- or AOA-related observations to obtain a final position estimate is usually referred to as *multilateration*. Whenever the number of observations reduces to three, then the process is referred to as *trilateration*, or *triangulation*, which correspond to classical geometric concepts.

Popular multilateration methods are usually divided into two main groups depending on the way the range- or AOA-related information is processed. Methods that find intersections between areas generated by means of the estimated range and/or AOA from several reference points to the NOI are referred to as *geometric multilateration*. On the other hand, methods that consider the minimization of range and/or AOA measurement errors via some statistical criterion are referred to as *statistical multilateration*.

In the presence of noise or measurement inaccuracies, a range or AOA-related observation between the i-th land reference and the NOI may be expressed as

$$r_i = f_i(x, y) + \eta_i, \tag{3.1}$$

where $f_i(x, y)$ is a function that describes the type of measurement (i.e., range- or AOA-related measurements) obtained between the i-th land reference and the NOI, (x, y) are the unknown NOI coordinates that we want to estimate, and η_i is the measurement error or noise that may be assumed to be a zero mean random variable. Function $f_i(x, y)$ may take the form of the Euclidean distance

$f_i(x, y) = \sqrt{(x - x_i)^2 + (y - y_i)^2}$ in the case of a range-related measurement, or the form $f_i(x, y) = \tan^{-1}\left[\frac{y-y_i}{x-x_i}\right]$ in the case of an AOA-related measurement.

Collecting N measurements arising from various land references in an observation vector $\mathbf{r} = [r_1 \ldots r_N]^H$, we obtain

$$\mathbf{r} = \mathbf{f}(x, y) + \boldsymbol{\eta}, \tag{3.2}$$

where $\mathbf{f}(x, y) = [f_1(x, y) \ldots f_N(x, y)]^H$ is a vector that collects the noise-free measurements between every land reference and the NOI (clearly this vector depends on the unknown NOI coordinates (x, y)), and $\boldsymbol{\eta} = [\eta_1 \ldots \eta_N]^H$ is a noise vector assumed to have a zero mean and covariance matrix $\mathbf{K} = E\{\boldsymbol{\eta}\boldsymbol{\eta}^H\}$. A diagonal covariance matrix $\mathbf{K}$ would imply uncorrelated measurements. The covariance between measurements will usually depend on the geographic separation of the land references. Measurements obtained from closely spaced land references will tend to be correlated due to the similarities of the propagation channel. The diagonal terms of the covariance matrix denote the variance of the noise terms and measure the reliability of the range- or AOA-related measurements.

Based on the observation model presented above, in the following sections we will describe several geometric and statistical multilateration techniques and discuss their performance and implementation issues.

3.2 GEOMETRIC MULTILATERATION

Range-based geometric multilateration techniques consist, in general, of the intersection of the areas generated by the estimated distances from several land references to the NOI. Such areas are centered at the land reference locations and extend in a radius defined by the estimated distance to the NOI. The distance between each reference point and the NOI can be estimated through several techniques discussed in Chapter 2 (TOA, AOA, RSS, etc.). On the other hand, AOA-based geometric multilateration consists of the intersection of lines of bearing whose slope corresponds to the AOA measurements between land references and the NOI.

Geometric techniques are commonly based on three range or AOA measurements, reducing the multilateration process to a trilateration one. When three reference points are connected to the NOI, the intersection of the generated areas can provide a good approximation of its position. Whenever more than three range or AOA measurements are available, it is a common practice to combine the results of several trilateration realizations obtained with different

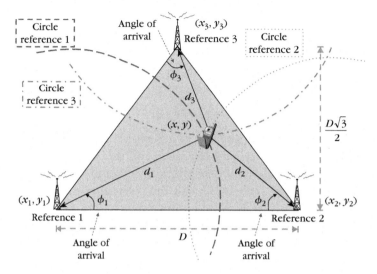

FIGURE 3.1

Fundamental multilateration technique. In a noise-free scenario, the point of intersection of circles (if TOA measurements are used), or ellipses (if TDOA measurements are used), centered at the land references with radii equal to the range measurements d_1, d_2, d_3 may be used to estimate the position of the NOI. Likewise, the point of intersection of lines of bearing drawn with slopes corresponding to AOA measurements ϕ_1, ϕ_2, ϕ_3 may also be used to localize the NOI. Estimation accuracy of the position of the NOI may be improved through a combination of intersections of ellipses, circles, and lines of bearing.

combinations of land-reference triplets either by averaging the multiple resulting position estimates or by finding their centroid.

Several geometric multilateration techniques are based on range estimation where a distance from the NOI to a fixed known point in the network is estimated according to measurements of TOA, TDOA, AOA or a combination of several of these. Techniques such as TDOA have been proposed as mobile positioning standards for 3G [1].

Assuming a planar scenario, the basic PL problem treated with geometric trilateration techniques is illustrated in Figure 3.1 for the case of three fixed land references with known position at points (x_i, y_i), $i = 1, 2, 3$, and an NOI whose unknown position (x, y) must be estimated. In this scenario, it is assumed that separation between any pair of land references is fixed to D meters. Estimated ranges or distances d_i, $i = 1, 2, 3$, and AOAs ϕ_i, $i = 1, 2, 3$, can be obtained from signal transmissions between the land references and the NOI, so that in a noise-free scenario, they satisfy the following equations:

$$d_i = \sqrt{(x - x_i)^2 + (y - y_i)^2} \tag{3.3}$$

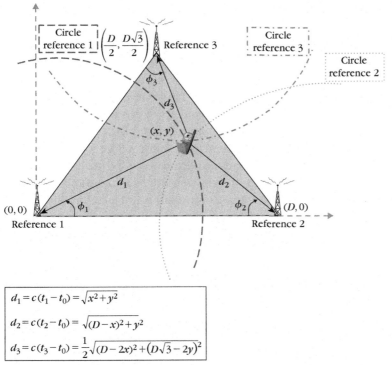

$$d_1 = c(t_1 - t_0) = \sqrt{x^2 + y^2}$$

$$d_2 = c(t_2 - t_0) = \sqrt{(D - x)^2 + y^2}$$

$$d_3 = c(t_3 - t_0) = \frac{1}{2}\sqrt{(D - 2x)^2 + (D\sqrt{3} - 2y)^2}$$

t_i: TOA at reference i; c: speed of light

FIGURE 3.2

TOA PL estimation technique.

$$\phi_i = \tan^{-1}\left[\frac{y - y_i}{x - x_i}\right]. \tag{3.4}$$

Without loss of generality, we can assume that the fixed land reference 1 is positioned at the origin of the coordinate system as shown in Figure 3.2. Depending on the parameter used to obtain information and the corresponding processing carried out to estimate distances, the location will be defined geometrically by the intersection of circles, hyperbolas, or lines of bearing as discussed in the following subsections.

3.2.1 Geometric Multilateration Based on TOA Measurements

To determine the range between the NOI and a land reference, the TOA technique exploits the knowledge of the propagation speed of a wave (acoustic or electromagnetic) in a specific medium such as air or water. In a 2D scenario, three land references are required to perform a trilateration process as illustrated in

Figure 3.2. If the estimated distance between the NOI and the first land reference is denoted as d_1, then the mobile will be located on a circle of radius d_1 centered at the reference coordinates. Note that the use of a single land reference would yield an NOI position ambiguity equivalent to the circumference of the aforementioned circle. This position ambiguity will be reduced to the area created by the intersection of two circles when another land reference is added to the system. Finally, in a noise-free scenario, a third land reference will allow the intersection of a third circle that will cause the position ambiguity to be reduced to a single point that should correspond to the true position of the NOI.

TOA-based trilateration methods require knowledge of absolute propagation times; thus, clock synchronization between the NOI and the land references is essential to avoid errors in distance computation and location estimates. Assuming that the NOI initiates transmission of a signal at time t_0, and that this signal is received at the i-th land reference at time t_i, the range or distance estimate may be found as

$$d_i = c\,(t_i - t_0),\tag{3.5}$$

where c is the wave propagation speed (equal to the speed of light $c = 3 \times 10^8$ m/s for the case of an electromagnetic wave propagating in the air). Ranges from three land references to the NOI are estimated and are used in the trilateration process, which, considering the scenario shown in Figure 3.2, may be formulated using the following system of nonlinear equations:

$$\begin{aligned}
d_1^2 &= x^2 + y^2,\\
d_2^2 &= (x_2 - x)^2 + (y_2 - y)^2 = (D - x)^2 + y^2,\\
d_3^2 &= (x_3 - x)^2 + (y_3 - y)^2 = \frac{1}{2}(D - 2x)^2 + \left(D\sqrt{3} - 2y\right)^2.
\end{aligned}\tag{3.6}$$

Note that this system of equations consists of three circles centered at each of the reference points. With this system of equations, we can find two different solutions for (x, y) that correspond to the two intersection points of the circles centered at their corresponding land references. The third equation in (3.6) allows us to solve the two-solution ambiguity by selecting the one that is closest to the range estimate given by the circle of radius d_3. One of the drawbacks of this method is that the third equation is not used in an optimal way to find the position location.

In a noise-free scenario, the three circles will intersect at exactly one point. Note, however, that in the case of noisy distance measurements, the circles will be displaced and their intersection will yield a polygon with an area that will correspond to a position ambiguity. It is common practice to use the final NOI position estimate as the centroid of this resulting polygon. Further, whenever more than three land references are available, one may obtain the NOI position estimate as the average, or centroid, of the estimated coordinates obtained from all possible trilateration realizations calculated with various triplet combinations of land references.

Although geometric trilateration procedures are simple to implement and rather computationally inexpensive, they are limited in the sense that they do not consider the reduction of noise effects and the optimal combination of multiple (more than three) range measurements. In the following sections, we will discuss basic techniques that help to overcome such problems by using statistical location-estimation methods such as least squares and maximum likelihood (ML), among others.

We can extend the TOA trilateration algorithm to the case of a 3D scenario as shown in Figure 3.3. In this scenario, each i-th reference with known position (x_i, y_i, z_i) has a range estimate based on TOA that defines a sphere of radius R_i around reference i within which the NOI must be located. The ambiguity, in this case, is defined not only by two points of intersection but by curves of intersection between two spheres. This ambiguity can be resolved by using the third range estimate obtained from the remaining land reference. The system of equations for this scenario is given by

$$R_i^2 = (X - x_i)^2 + \left(Y - y_i\right)^2 + (Z - z_i)^2, \quad i = 1, 2, 3. \tag{3.7}$$

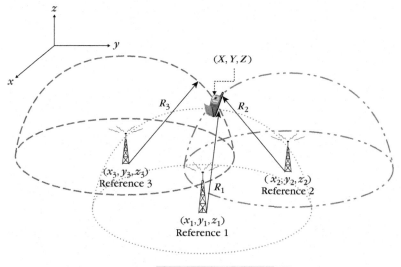

$$R_i = ct_i = \sqrt{(x_i - X)^2 + (y_i - Y)^2 (z_i - Z)^2}$$

t_i: TOA at reference i; c: speed of light

FIGURE 3.3

TOA in 3D.

We can see that each of the equations in (3.7) is a sphere of radius R_i, $i = 1, 2, 3$, centered at each of the references.

3.2.2 Geometric Multilateration Based on AOA Measurements

The AOA technique estimates the position location of an NOI by means of angular direction observations measured with respect to a reference axis using directional antennas or antenna arrays as shown in Figure 3.4. This direction, as it has been explained in Chapter 2, may be calculated through the phase differences of the elements of an antenna array.

At least two AOA measurements from two different references are necessary to provide the position location estimation of a mobile, one less than the number required for the TOA localization case. The system of equations for this scenario is given by

$$x = d_i \cos(\phi_i) + x_i,$$

$$y = d_i \sin(\phi_i) + y_i, \quad i = 1, 2, 3.$$

(3.8)

A single AOA measurement restricts the location of the source along a line in the estimated line of bearing [22]. When multiple AOA measurements are

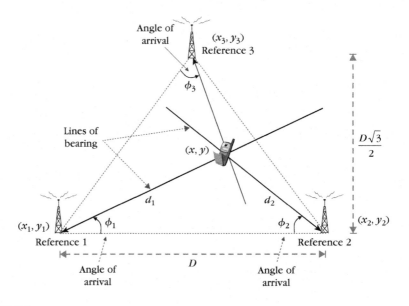

FIGURE 3.4

AOA technique.

obtained simultaneously by multiple land reference stations, a triangulation method may be used to form a location estimate of the NOI at the intersection of these lines [23]. In theory, direction-finding systems require only two receiving sensors to locate a mobile user, but in practice, and to improve accuracy, finite angular resolution, multipath, and noise often dictate the need for more than two references.

An advantage of AOA-based techniques is that they do not require clock synchronization; however, accurate angle measurements may require directional antennas, multiple-element antenna arrays, and possibly computationally expensive array processing algorithms.

3.2.3 Geometric Multilateration Based on TDOA Measurements

A variant of the TOA technique is the TDOA, which uses differences in the TOAs to locate the mobile. The TDOA involves the intersection of hyperbolas rather than circles as shown in Figure 3.5. The main advantage of TDOA over TOA-based methods is that the former do not require knowledge of the time at which a transmission took place at the NOI (i.e., t_0). Therefore strict time synchronization between an NOI and land references is not required. Note, however, that time synchronization among the different land references is still necessary.

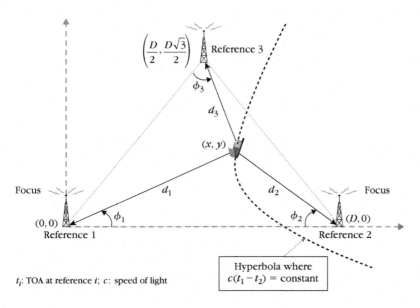

t_i: TOA at reference i; c: speed of light

FIGURE 3.5

TDOA technique.

The TDOA between land references i and j (i.e., $t_i - t_j$) may be used to obtain the distance difference, which may be written as

$$d_{ij} = d_i - d_j = c\,(t_i - t_0) - c\,(t_j - t_0) = c\,(t_i - t_j), \quad i = 1,\,2,\,3,\,j = 1,\,2,\,3,\,i \neq j. \quad \textbf{(3.9)}$$

Notice that the common time reference t_0 (i.e., the time at which the NOI started a transmission) cancels out in this calculation and hence, as it has been stated previously, its knowledge is not necessary. It follows from Equation (3.9) that

$$d_i = d_{ij} + d_j; \quad \textbf{(3.10)}$$

thus, it is possible to calculate the square of the range estimate seen by reference 2 as

$$
\begin{aligned}
d_2^2 &= (d_{21} + d_1)^2 \\
&= (x_2 - x)^2 + (y_2 - y)^2 \\
&= x_2^2 - 2x_2 x + x^2 + y_2^2 - 2y_2 y + y^2 \\
&= x_2^2 - 2x_2 x + y_2^2 - 2y_2 y + d_1^2,
\end{aligned}
\quad \textbf{(3.11)}
$$

where the last equality follows from the fact that reference 1 is assumed to be at the coordinate system's origin $(0, 0)$, such that

$$d_1^2 = x^2 + y^2. \quad \textbf{(3.12)}$$

We can rearrange terms in (3.12) to obtain

$$\left(d_{21}^2 - x_2^2 - y_2^2\right) + 2d_{21}d_1 = -2x_2 x - 2y_2 y, \quad \textbf{(3.13)}$$

and we can extend this procedure to obtain another equation as

$$\left(d_{31}^2 - x_3^2 - y_3^2\right) + 2d_{31}d_1 = -2x_3 x - 2y_3 y. \quad \textbf{(3.14)}$$

Since, except for the NOI coordinates, (x, y), all the terms in the equations derived above are known, we can solve the system of equations to obtain the unknown NOI position.

Another variation of TOA-based techniques is the time sum of arrival (TSOA) of the propagating signal from an NOI to two land references to produce a range sum measurement. The range sum estimate defines an ellipsoid with foci at two receivers, and when multiple range sum measurements are obtained, the position location estimate of the user occurs at the intersection of the ellipsoids as shown in Figure 3.6. Consequently, range sum position location systems are also known as TSOA or elliptical position location systems. When the transmitted power at the mobile is known, the receiver can measure the signal strength

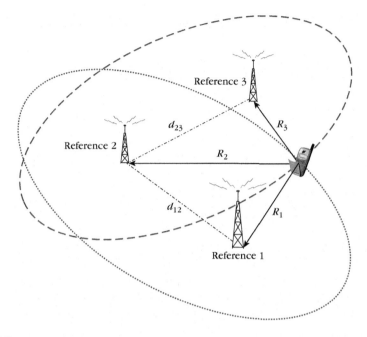

FIGURE 3.6

Technique using ellipsoids.

and provide a distance estimation between the mobile and the receiver using mathematical models of path loss that depend on distance.

3.3 STATISTICAL MULTILATERATION

In previous sections, we showed that geometric multilateration techniques rely on the assumption of noise-free measurements, and thus are strongly affected by noisy range or AOA observations. Position estimation ambiguities caused by noise effects can only be conciliated through heuristic methods such as finding the centroid of the resulting areas of uncertainty. Further, these techniques do not provide an optimal way to combine more than three observations that may well be needed in order to improve accuracy in noisy scenarios.

In what follows, we will present some statistical multilateration procedures that allow the optimal combination of three or more range- or AOA-related measurements in order to obtain a position estimate of an NOI while minimizing the effects of observation noise. Even measurements arising from different types of observations such as TOA, TDOA, and AOA may be combined into a single set of equations that may be solved optimally in order to reduce the effects of noise.

Consider the observation of N noisy measurements collected in the vectorial observation model described in Equation (3.2) and rewritten here for convenience:

$$\mathbf{r} = \mathbf{f}(x, y) + \boldsymbol{\eta}. \tag{3.15}$$

The problem at hand is to combine all N observations in an optimal way such that the noise effects are minimized. In the following, we will describe popular optimal statistical multilateration procedures commonly used for solving the PL problem.

3.3.1 Least-Squares Multilateration

Let us start by defining vector $\mathbf{q}$ as a bidimensional column vector containing the NOI unknown coordinates (x, y) such that $\mathbf{q} = [xy]^H$. Then, given a set of N noisy observations $\mathbf{r} = \mathbf{f}(\mathbf{q}) + \boldsymbol{\eta}$, the least-squares (LS) optimization problem may be posed as follows:

$$\hat{\mathbf{q}} = \arg\min_{\mathbf{q}} [\mathbf{r} - \mathbf{f}(\mathbf{q})]^H [\mathbf{r} - \mathbf{f}(\mathbf{q})]. \tag{3.16}$$

Whenever the noise covariance matrix $\mathbf{K}$ is known, a variant of the LS problem called the weighted LS (WLS) problem [23] may be posed as

$$\hat{\mathbf{q}} = \arg\min_{\mathbf{q}} [\mathbf{r} - \mathbf{f}(\mathbf{q})]^H \mathbf{K}^{-1} [\mathbf{r} - \mathbf{f}(\mathbf{q})], \tag{3.17}$$

where the cost function is defined as

$$J = [\mathbf{r} - \mathbf{f}(\mathbf{q})]^H \mathbf{K}^{-1} [\mathbf{r} - \mathbf{f}(\mathbf{q})]. \tag{3.18}$$

In this last problem, the inverse covariance matrix $\mathbf{K}^{-1}$ acts as a weighting factor that emphasizes measurements containing the smallest noise variances.

Clearly, problems described in Equations 3.16 and 3.17 will be nonlinear since $f_i(\mathbf{q})$, $i = 1, \ldots, N$ may take the form of the nonlinear Euclidean distance or line-of-bearing angle equations as they appear in Equations (3.3), and (3.4), respectively. These nonlinear LS problems may be solved by numerous numeric minimization techniques such as the interior-reflective Newton algorithm described in Coleman and Li [2]. Solution of the nonlinear LS problem may pose several problems, such as high computational loads and, more importantly, convergence issues caused by local minimum points. To avoid these issues, a linearization procedure, based on a first-order Taylor series expansion of function vector $\mathbf{f}(\mathbf{q})$ around a known position vector $\mathbf{q_0} = [x_0 y_0]^H$, may be applied

[4, 25]. The result of this procedure will be a linearized LS (LLS) problem whose solution is well known and straightforward to obtain [23]. Let us then present the derivation of the LLS algorithm.

It is easy to show that the first-order Taylor series expansion of the function vector $\mathbf{f}(\mathbf{q})$ around the known point $\mathbf{q}_0 = [x_0 y_0]^H$ (it will be assumed that the value $\mathbf{q}_0$ is sufficiently close to the true position vector $\mathbf{q}$ so that the series expansion is an accurate approximation) is given by

$$\mathbf{f}_l(\mathbf{q}) = \mathbf{f}(\mathbf{q}_0) + \mathbf{D}(\mathbf{q} - \mathbf{q}_0), \qquad (3.19)$$

where subindex l in term $\mathbf{f}_l(\mathbf{q})$ stands for the *linearized* approximation, and $\mathbf{D}$ is a $N \times 2$ matrix given by

$$\mathbf{D} = \begin{bmatrix} \frac{\partial f_1(\mathbf{q})}{\partial x} \big|_{\mathbf{q}=\mathbf{q}_0} & \frac{\partial f_1(\mathbf{q})}{\partial y} \big|_{\mathbf{q}=\mathbf{q}_0} \\ \vdots & \vdots \\ \frac{\partial f_N(\mathbf{q})}{\partial x} \big|_{\mathbf{q}=\mathbf{q}_0} & \frac{\partial f_N(\mathbf{q})}{\partial y} \big|_{\mathbf{q}=\mathbf{q}_0} \end{bmatrix}. \qquad (3.20)$$

For instance, if all functions in $\mathbf{f}(\mathbf{q})$ correspond to Euclidean distances between land references with known positions (x_i, y_i), $i = 1 \dots N$, and a NOI, terms in matrix D become

$$\frac{\partial f_i(\mathbf{q})}{\partial x}\big|_{\mathbf{q}=\mathbf{q}_0} = \frac{x_0 - x_i}{\left[(x_i - x_0)^2 + (y_i - y_0)^2\right]^{1/2}}, \quad i = 1 \dots N, \qquad (3.21)$$

and

$$\frac{\partial f_i(\mathbf{q})}{\partial y}\big|_{\mathbf{q}=\mathbf{q}_0} = \frac{y_0 - y_i}{\left[(x_i - x_0)^2 + (y_i - y_0)^2\right]^{1/2}}, \quad i = 1 \dots N. \qquad (3.22)$$

Substituting the linearized function vector given in Equation (3.19) into the cost function in Equation (3.18), and redefining the observation vector as

$$\mathbf{r}' = \mathbf{r} - \mathbf{f}(\mathbf{q}_0) + \mathbf{D}\mathbf{q}_0, \qquad (3.23)$$

the cost function becomes

$$J = (\mathbf{r}' - \mathbf{D}\mathbf{q})^H \mathbf{K}^{-1}(\mathbf{r}' - \mathbf{D}\mathbf{q}), \qquad (3.24)$$

which is now clearly linear.

Minimizing (3.24) with respect to $\mathbf{q}$ yields the linearized WLS estimator, which is given by [23] as

$$\hat{\mathbf{q}} = (\mathbf{D}^H \mathbf{K}^{-1} \mathbf{D})^{-1} \mathbf{D}^H \mathbf{K}^{-1} \mathbf{r}', \qquad (3.25)$$

where the hat superscript denotes an estimated quantity.

Expanding $\mathbf{r}'$ as in (3.23) and rearranging terms, the estimator can be rewritten as

$$\hat{\mathbf{q}} = \mathbf{q}_0 + (\mathbf{D}^H \mathbf{K}^{-1} \mathbf{D})^{-1} \mathbf{D}^H \mathbf{K}^{-1} [\mathbf{r} - \mathbf{f}(\mathbf{q}_0)]. \qquad (3.26)$$

This last equation suggests the possibility of finding the WLS estimate $\hat{\mathbf{q}}$ with a recursion that uses, at each step, the actual estimate of $\hat{\mathbf{q}}$ as the value of $\hat{\mathbf{q}}_0$. This recursion is given as

$$\hat{\mathbf{q}}(n+1) = \hat{\mathbf{q}}(n) + (\mathbf{D}^H\mathbf{K}^{-1}\mathbf{D})^{-1}\mathbf{D}^H\mathbf{K}^{-1}[\mathbf{r} - \mathbf{f}(\hat{\mathbf{q}}(n))]. \qquad (3.27)$$

It has been observed that this recursion converges even when the initial $\hat{\mathbf{q}}(0)$ value is far from the true position vector $\mathbf{q}$ [17].

Finally, it is easy to show, after some algebraic manipulations, that the bias and covariance of the WLS estimator are given respectively by

$$\mathbf{b} = E\{\hat{\mathbf{q}}\} - \mathbf{q} = (\mathbf{D}^H\mathbf{K}^{-1}\mathbf{D})^{-1}\mathbf{D}^H\mathbf{K}^{-1}[\mathbf{f}(\mathbf{q}) - \mathbf{f}_l(\mathbf{q})], \qquad (3.28)$$

and

$$\mathbf{C} = E\{[\hat{\mathbf{q}} - E\{\hat{\mathbf{q}}\}][\hat{\mathbf{q}} - E\{\hat{\mathbf{q}}\}]^H\} = (\mathbf{D}^H\mathbf{K}^{-1}\mathbf{D})^{-1}, \qquad (3.29)$$

where $\mathbf{f}_l(\mathbf{q})$ is the linearized function vector given in (3.19). Clearly, the bias is nonzero due to the linearization error $\mathbf{f}(\mathbf{q}) - \mathbf{f}_l(\mathbf{q})$. Also, the covariance of the estimator is dependent on both the covariance matrix of the distance observations $\mathbf{K}$ and the linearization matrix $\mathbf{D}$. Note that the covariance expression $(\mathbf{D}^H\mathbf{K}^{-1}\mathbf{D})^{-1}$ appears in the recursion in Equation (3.27). This allows us to compute both the position estimate and its covariance simultaneously.

Up to this point no assumptions have been made about the probability distribution of noise vector $\boldsymbol{\eta}$. Note, however, that if this vector is Gaussian distributed, then the estimator given by (3.25) corresponds to the maximum likelihood estimator.

3.3.2 LS Multilateration with Uncertain Reference Node Positions

Up to this point, we have assumed perfect knowledge of the position coordinates of every land reference in the system. This condition will hardly be met in a real system due to measurement and mapping errors. If not taken into consideration, land reference position uncertainties may strongly degrade performance of the PL techniques.

A modified LWLS position estimation algorithm may be used to account for land reference position uncertainties. This algorithm was presented in Kovavisaruch and Ho [9], and is summarized in this section.

Let us denote the set of true position coordinates for the i-th land reference as $\mathbf{q}_i = [x_i \, y_i]^H$, $i = 1 \ldots N$ (vector $\mathbf{q} = [\mathbf{x} \, \mathbf{y}]^H$ without any subindex still corresponds to the unknown position coordinates of the NOI), and the corresponding set of uncertain land reference position observations as

$$\tilde{\mathbf{q}}_i = \mathbf{q}_i + \boldsymbol{\Delta}_i = [\tilde{x}_i \, \tilde{y}_i]^H = [x_i + \Delta_{x_i} \, y_i + \Delta_{y_i}]^H, \quad i = 1 \ldots N, \qquad (3.30)$$

where Δ_{x_i} and Δ_{y_i} are coordinate error terms that may be grouped into vector $\mathbf{\Delta}_i = [\Delta_{x_i} \; \Delta_{y_i}]^H$, $i = 1 \ldots N$. We may then define a $3N$-dimensional extended vector of observations as

$$\tilde{\mathbf{s}} = [\mathbf{r}^H \; \tilde{\mathbf{q}}_1^H \ldots \tilde{\mathbf{q}}_N^H]^H, \tag{3.31}$$

where $\mathbf{r}$ corresponds to the N noisy range- or AOA-related observations as described in Equation (3.15), and $\tilde{\mathbf{q}}_i$, $i = 1 \ldots N$ are the available erroneous land reference position measurements. Equation (3.31) may be written in extended form as

$$\begin{bmatrix} \mathbf{r} \\ \tilde{\mathbf{q}}_1 \\ \vdots \\ \tilde{\mathbf{q}}_N \end{bmatrix} = \begin{bmatrix} \mathbf{f}(\mathbf{q}) \\ \mathbf{q}_1 \\ \vdots \\ \mathbf{q}_N \end{bmatrix} + \begin{bmatrix} \boldsymbol{\eta} \\ \mathbf{\Delta}_1 \\ \vdots \\ \mathbf{\Delta}_N \end{bmatrix}, \tag{3.32}$$

or simply

$$\tilde{\mathbf{s}} = \mathbf{s} + \mathbf{n}, \tag{3.33}$$

where $\mathbf{s} = [\mathbf{f}(\mathbf{q})^H \; \mathbf{q}_1^H \ldots \mathbf{q}_N^H]^H$, and $\mathbf{n} = [\boldsymbol{\eta}^H \; \mathbf{\Delta}_1^H \ldots \mathbf{\Delta}_N^H]$.

We may assume that the extended noise vector $\mathbf{n}$ has zero mean and that the range- or AOA-related observation noise vector $\boldsymbol{\eta}$ and the land reference position error vectors $\mathbf{\Delta}_i$, $i = 1 \ldots N$ are statistically independent. Thus, the extended noise vector covariance matrix may be written as

$$\mathbf{Q} = E\{\mathbf{n}\mathbf{n}^H\} \begin{bmatrix} \mathbf{K} & \mathbf{0}^H \\ \mathbf{0} & \mathbf{K}_\Delta \end{bmatrix}, \tag{3.34}$$

where $\mathbf{0}$ corresponds to a $(2N \times N)$ matrix of zeros, $\mathbf{K}$ is the $(N \times N)$-dimensional covariance matrix of noise vector $\boldsymbol{\eta}$ as described in Section 3.1, and $\mathbf{K}_\Delta$ is the $(2N \times 2N)$-dimensional covariance matrix of the overall vector of land reference position errors $[\mathbf{\Delta}_1^H \ldots \mathbf{\Delta}_N^H]^H$.

If we treat the true NOI, and land reference coordinates as unknown parameters and group them in vector $\boldsymbol{\theta}$ such that $\boldsymbol{\theta} = [x \; y \; x_1 \; y_1 \ldots x_N \; y_N]^H$, then we may pose the following WLS problem to estimate this parameter vector as follows:

$$\hat{\boldsymbol{\theta}} = \arg \min_{\boldsymbol{\theta}} [\tilde{\mathbf{s}} - \mathbf{s}(\boldsymbol{\theta})]^H \mathbf{Q}^{-1} [\tilde{\mathbf{s}} - \mathbf{s}(\boldsymbol{\theta})], \tag{3.35}$$

where we have now emphasized the dependence of the noiseless observation vector on parameter $\boldsymbol{\theta}$. Again, we note that this is a nonlinear least-squares problem whose solution may involve large computational loads and convergence problems due to local minima. A linearized WLS problem may, however, be derived following the same Taylor series expansion procedure used in Section 3.3.1. Let us then approximate the noiseless observation vector $\mathbf{s}$ using its first-order Taylor series expansion around a known parameter vector $\boldsymbol{\theta}_0$ so that

$$s_l(\boldsymbol{\theta}) = s(\boldsymbol{\theta}_0) + \mathbf{D}(\boldsymbol{\theta} - \boldsymbol{\theta}_0), \tag{3.36}$$

where now the linearization matrix $\mathbf{D}$ corresponds to

$$
\mathbf{D} = \left.\begin{bmatrix}
\frac{\partial \mathbf{f(q)}}{\partial \mathbf{q}} & \frac{\partial \mathbf{f(q)}}{\partial \mathbf{q}_1} & \cdots & \frac{\partial \mathbf{f(q)}}{\partial \mathbf{q}_N} \\
\frac{\partial \mathbf{q}_1}{\partial \mathbf{q}} & \frac{\partial \mathbf{q}_1}{\partial \mathbf{q}_1} & \cdots & \frac{\partial \mathbf{q}_1}{\partial \mathbf{q}_N} \\
\vdots & \vdots & \vdots & \vdots \\
\frac{\partial \mathbf{q}_N}{\partial \mathbf{q}} & \frac{\partial \mathbf{q}_N}{\partial \mathbf{q}_1} & \cdots & \frac{\partial \mathbf{q}_N}{\partial \mathbf{q}_N}
\end{bmatrix}\right|_{\boldsymbol{\theta} = \boldsymbol{\theta}_0},
\tag{3.37}
$$

and where we note that $\frac{\partial \mathbf{q}_i}{\partial \mathbf{q}} = [\mathbf{0}]_{2 \times 2}$, $\frac{\partial \mathbf{q}_i}{\partial \mathbf{q}_j} = [\mathbf{0}]_{2 \times 2}$ for $i \neq j$, and $\frac{\partial \mathbf{q}_i}{\partial \mathbf{q}_i} = [\mathbf{I}]_{2 \times 2}$. Finally, following the same procedure as in Section 3.3.1, the estimate of vector $\boldsymbol{\theta}$ may be found recursively as follows:

$$
\hat{\boldsymbol{\theta}}(n+1) = \hat{\boldsymbol{\theta}}(n) + (\mathbf{D}^H \mathbf{Q}^{-1} \mathbf{D})^{-1} \mathbf{D}^H \mathbf{Q}^{-1} [\tilde{\mathbf{s}} - \mathbf{s}(\hat{\boldsymbol{\theta}}(n))].
\tag{3.38}
$$

Note that apart from estimating the unknown NOI coordinates, this algorithm also estimates the land reference positions that were originally known with uncertainty. The overall effect of this procedure is to obtain better NOI position estimates while also obtaining corrections for the erroneous land reference coordinates.

3.3.3 Hybrid Location Estimation Systems

The WLS criterion does not pose any condition on the type of equations that may be used in the minimization problem described in Equation (3.17). This means that we may combine different types of measurements to obtain position estimates. For instance, a subset of the N equations may correspond to AOA observations, while another subset may correspond to range observations obtained from TOA measurements, and another subset may correspond to range difference observations obtained from TDOA measurements. Location estimation systems that combine different types of observations are commonly referred to as *hybrid* localization systems.

3.4 LOCATION ESTIMATION IN MULTIHOP SCENARIOS

As stated in the previous section, geographic and statistical multilateration techniques rely on the estimation of range between a node with unknown location and a set of land reference nodes whose location is fixed and known. These techniques assume that direct signal transmissions between the land references and the NOI can be established at any time so that TOA, TDOA, AOA, or RSS observations are available for the estimation of range and/or line of bearing. GPS and cellular localization systems are examples of schemes that apply direct multilateration to estimate the position of a device or mobile phone, respectively. Algorithms that relay on direct connectivity between land references and an NOI are usually referred to as single-hop schemes.

In the case of ad hoc and mobile sensor-network scenarios, a mobile node whose location must be estimated may not have enough transmission power to establish a direct connection to any or some land references in the network, and the direct multilateration techniques treated in the previous sections of this chapter will not be applicable. In this type of environments, two nodes that are not within the maximum transmission radius of each other communicate by means of a concatenation of multiple links between intermediate neighboring nodes. Since a multilateration process cannot be applied directly in an ad hoc environment due to the lack of direct connectivity of nodes to well-located land references, multihop algorithms are needed.

Iterative algorithms are a popular type of multihop position-estimation scheme [13, 15]. In these schemes, any type of useful data, such as land reference coordinates and/or measurements of proximity, range, and AOA between network nodes, is propagated iteratively throughout one or more sectors of the network so that, after a number of iterations, an NOI that was initially away from the neighborhood of any land reference nodes may receive this information and use it to estimate its position. During the iteration process, intermediate nodes may also be able to estimate their positions and may act as newly acquired anchor nodes if desired. Depending on the size of the network sector and the number of intermediate nodes involved in the data propagation process, iterative algorithms may be classified as localized or nonlocalized [13].

In a typical nonlocalized iterative algorithm, a node that is in the vicinity (within its transmission reachability radius) of more than one anchor node may be able to estimate its position using multilateration techniques based on (possibly combined) measurements of proximity, range, and AOA. When this node accomplishes its position estimation procedure, it may now act as a new anchor node (with reduced position accuracy) and assist its neighboring nodes in their own position estimation processes. Newly generated anchors are iteratively propagated through the network until all nodes locate themselves or until sufficient information reaches an NOI so that it can be localized. In order to request the location of an NOI at a specific time when this node is not in the neighborhood of a sufficient number of anchor nodes, several intermediate nodes may need to be located throughout the network in order to propagate the necessary anchors into the NOI's reachability region. The number of necessary intermediate nodes that need to be involved in the iterative process in order to reach an NOI will depend on the network density, the nodes' reachability radii, and the position and density of original land references. Depending on these parameters, and on the position estimation algorithm itself, the region occupied by the intermediate nodes may expand a large sector or even the complete network area. Nonlocalized methods are highly sensitive to anchor densities and can get stuck in places where these densities are sparse. Further, error propagation of newly created anchor nodes may become an issue in large network scenarios.

Localized iterative methods, on the other hand, do not require the NOI to be at one-hop proximity to anchor nodes. The only requirement is that the NOI has connectivity, via multihop routes, to three or more land references. Hence, the only intermediate nodes involved in the NOI position estimation process are the nodes that forms those routes. If shortest-distance routing is used in the network, then it is clear that this type of algorithm will only involve intermediate nodes located in limited network regions that lie in the shortest paths between land references and the NOI. It is important to note that these algorithms will not necessarily require that intermediate nodes become anchors (i.e., there is no need to estimate their coordinates) as was the case in nonlocalized techniques. For these reasons, localized algorithms considerably reduce computational load, power consumption, communications traffic, and the time needed to obtain a position estimate of a specific single NOI.

Some nonlocalized methods are based on multilateration of propagated anchor nodes and on the solution of computationally intensive global nonlinear optimization problems to reduce error propagation effects [3, 20, 21]. Other methods may involve obtaining pairwise range estimates between every node in the network in order to solve an ML problem that estimates the position of all network nodes at once [14]. Simpler methods may involve purely geometric range-free solutions such as the centroid localization algorithm [1] where the NOI estimates its position as the centroid of neighboring anchor nodes, or the approximate point-in-triangulation (APIT) algorithm [6] where anchor nodes are generated (and then propagated through the network until the NOI is reached) by calculating the maximum overlap of triangular regions formed by triplets of neighboring anchor nodes. Other simple nonlocalized iterative position estimation methods called Euclidean and DV-bearing (DV stands for distance vector) are presented in Niculescu and Nath [11,12]. These schemes require an NOI to be in the proximity of at least two neighboring nodes that know their distance to a land reference. The Euclidean method is based on ranging while the DV-bearing method is based on ranging and AOA measurements. APIT, Euclidean, and DV-bearing algorithms do not take into consideration error propagation of newly created beacons, and may be considerably affected by shadowing and small-scale fading, as has been posed in Zhou et al. [5] and He et al. [6].

A well-known, localized iterative position estimation scheme is the ad hoc positioning system (APS) algorithm [11]. APS has two variations called DV-hop and DV-distance. DV-hop uses mere connectivity to infer range estimates of an NOI to several land references. It employs two stages, a calibration stage in which the average size of a hop is estimated by a land reference and distributed to nearby nodes, and a second stage in which an NOI measures the number of hops required to reach different land references via the shortest possible paths. Land references know their exact coordinates and the exact coordinates of all other land references in the network so they are able to estimate average hop sizes by communicating constantly with each other and measuring the number of hops required for a message to travel the distance between them. Distance in hops from an NOI to a land reference is then multiplied by the estimated size of a hop to produce a range estimate. If the NOI is able to obtain range estimates to three or more land references, then it can perform multilateration to find its position estimate. A variation of DV-hop is the DV-distance algorithm where the NOI uses measured distances between nodes at each hop in a shortest-path connection to a land reference and adds these distances together to obtain an overestimate of its range to that land reference. This overestimation is corrected by applying a correcting factor whose value is continuously being broadcasted by the land references in the network. Similar to the DV-hop case, the correcting factor is estimated by the land references in the network while they communicate with each other and compare resultant hop-distance sums to actual geographic distances.

Disadvantages of these relatively simple methods are that they require a uniform geographic distribution of nodes, nodes with similar transmission power and antenna gains, and a large node density in order for hop sizes and number of hops per unit distance to be fairly constant throughout the network. These methods do not take into consideration measurement error statistics, and their performance may degrade considerably in the presence of estimation noise. These algorithms are very sensitive to channel variations due to the fact that the hop-size estimation becomes less precise in the presence of irregular radio reception patterns. Examples of this behavior have been presented in [6]. Finally, because DV-hop–based algorithms use flood-based correction factor propagation, they generate large communications overhead and power consumption.

To overcome the limitations of the DV-hop and DV-distance schemes, a localized iterative technique based on a dead-reckoning–like scheme has been proposed in Hernandez et al. [7]. In this scheme, error accumulation effects are accounted for by modeling of the statistical behavior of ranging and AOA noisy measurements at each hop. As in DV-hop and DV-distance, this algorithm does not require the NOI to be at a one-hop proximity from at least three anchors. Instead, the requirement consists of at least three land references being able to establish a link with any NOI via multiple hops.

Multihop algorithms may also be classified as centralized or distributed [7, 13]. This category division has direct impact on the applicability of the algorithm and on the efficiency of the positioning system. Centralized algorithms collect measurements of the entire network at a central node that optimizes for the position estimation of the NOI. Besides gathering measurements of the entire network in a single main powerful central node that must deal with large and complex data structures, these algorithms may face the problem of communication bottleneck and higher energy drain at and near the central node caused by the large amount of traffic associated with large networks.

Distributed algorithms require no specialized central node because nodes share information with their neighbors in an iterative fashion. These algorithms rely on the propagation of anchor nodes in the network until the NOI ends up being surrounded by them. These algorithms have the advantage of load balancing because they do not depend on a single central node that represents a single failure point, where, if this node fails, the entire chain breaks down and no localization can be achieved. As they iteratively share information among themselves, nodes must be capable of computing data and handling the necessary calculations.

In the following sections, we will describe some of the aforementioned multihop localization techniques. Most of them rely on the multilateration problem described in Section 3.1, but some use another type of heuristic to obtain position estimates.

3.4.1 Centroid Algorithm

Position estimation algorithms that do not use any type of signal measurement to infer range or AOA information between a land reference and an NOI are usually referred to as *range-free* schemes. The centroid algorithm, proposed in Bulusu et al. [1], falls into this category.

Consider an NOI that is in the vicinity of N anchor nodes with coordinates $(x_i, y_i), i = 1 \ldots N$. The anchor nodes communicate and transmit their coordinate points to the NOI. After receiving the anchor coordinates, the NOI simply estimates its location as the centroid of those points. The centroid of all anchor coordinates is calculated as follows:

$$(\hat{x}, \hat{y}) = \left(\frac{1}{N} \sum_{i=1}^{N} x_i, \ \frac{1}{N} \sum_{i=1}^{N} y_i \right). \tag{3.39}$$

The centroid algorithm is a nonlocalized distributed scheme where an NOI needs to be in the vicinity of N anchor nodes. Iterative anchor node propagation may be necessary if, initially, no anchor nodes exist in the vicinity of the NOI. Note that the centroid algorithm may also be applicable to single-hop scenarios, but the accuracy will degrade as the reachability radii of the network devices increases.

Although this algorithm may suffer from limited accuracy, it has the advantage of being very simple and easy to implement.

3.4.2 Approximate Point-in-Triangulation Algorithm

APIT [6] is a nonlocalized iterative algorithm that uses beacon transmissions from anchor nodes. It employs an area-based approach to perform location estimation by isolating the environment into triangular regions between beaconing nodes as shown in Figure 3.7. A node's presence inside or outside these triangular regions allows this node to narrow down the area in which it can potentially reside. By using combinations of anchor positions, the diameter of the estimated area in which a node resides can be reduced to provide accurate location estimates.

The theoretical method used to narrow down the possible area in which a node resides is called the point-in-triangulation (PIT) test. In this test, a node chooses three anchors from all audible anchors and tests whether it is inside the triangle formed by connecting these three anchors. APIT repeats this PIT test with different audible anchor triplets until all combinations are exhausted or the required accuracy has been achieved. In the final step, APIT calculates the center of gravity (COG) of the intersection of all the triangles in which a node resides to determine its estimated position. Then the APIT algorithm may be summarized in four steps: (1) beacon exchange, (2) PIT testing, (3) APIT aggregation, and (4) COG calculation.

The critical component of the APIT scheme is to determine whether an NOI is inside a triangle formed by three anchor nodes. Theoretically, if an NOI is inside a triangle formed by anchor nodes *A*, *B*, *C*, as shown in Figure 3.8,

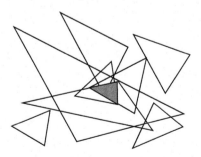

FIGURE 3.7

APIT is based on triangular intersections.

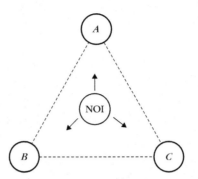

FIGURE 3.8

PIT test.

when this node is shifted in any direction, the new position must be nearer to (or further from) at least one anchor A, B, or C. On the other hand, if the NOI is outside the ABC triangle, when the NOI is shifted, there must exist a direction in which the position of this NOI is further or closer to all three anchors A, B, C simultaneously. Although theoretically correct, in practice it is impossible to recognize the directions of departure of the NOI without its actual movement. Further, exhaustively testing all possible directions in which the NOI might depart or approach nodes A, B, and C simultaneously is impossible. For these reasons, an approximate PIT test must be implemented in practice.

The approximate PIT test relies on the assumption that in a certain propagation direction, defined to be within a narrow angle from the transmitting anchor, the received signal strength decreases monotonically with distance. Approximate PIT uses neighbor information, exchanged via beaconing, to emulate the node movement required in the theoretical test. Referring to Figure 3.9, if no neighbor of the NOI is further from/closer to all three anchors A, B, C simultaneously, the NOI assumes that it is inside the triangle ABC. Otherwise, the NOI assumes that it resides outside this triangle. Although this is a powerful test, it may suffer from errors in cases when the NOI resides near the edge of the triangle, or when the NOI is placed in the very improbable position where the simultaneous further from/closer to condition in relation to all three anchors A, B, C does not hold.

Once an individual PIT test is finished, APIT aggregates the results (inside/outside decisions) by means of the grid shown in Figure 3.10. The network area is divided into small square regions on a grid and this grid array is used to represent the maximum area in which a node will likely reside. For each PIT

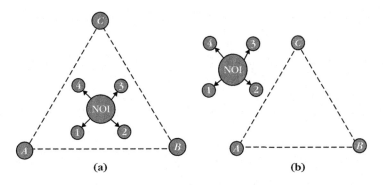

FIGURE 3.9

Approximate PIT test. (a) Inside case. (b) Outside case.

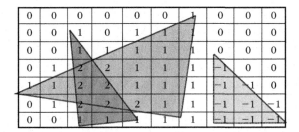

FIGURE 3.10

Aggregation procedure.

inside decision, the values of the grid regions over which the corresponding triangles reside are incremented. On the other hand, for an *outside* decision, the grid area is decremented. After all triangular regions have been computed, the resulting information is used to find the maximum overlapping area (i.e., the grid squares with the largest counters equal to two in the example figure), which is then used to calculate the center of gravity for position estimation.

Although simple to implement, APIT may suffer from irregular node transmission patterns, and from large- and small-scale fading that may prevent received signal strength from reducing monotonically with distance, in which case PIT test failure rates will increase. Further, APIT accuracy will be strongly dependent on the grid size used in the aggregation step of the algorithm. Finally, APIT is very sensitive to the network's anchor node density. Very few anchor nodes will not allow enough triangular regions to overlap and, in this scenario, the accuracy of the algorithm will decrease.

3.4.3 Ad Hoc Positioning System Algorithms

As the name implies, ad hoc positioning system (APS) algorithms were specifically designed for ad hoc scenarios. They consist of relatively simple multihop schemes that obtain node position estimates based on indirect measurements of distance [11] or AOA [12] between anchor nodes and NOIs. In this section, we will describe four APS modalities called Euclidean, DV-bearing, DV-hop, and DV-distance. Here, the acronym *DV* stands for distance vector, which in turn refers to the class of algorithms that rely on distance observations based on calculations of the magnitude of vectorial sums obtained from range and/or AOA measurements at each hop of a multihop link connecting an anchor node to an NOI.

Euclidean APS

This is a nonlocalized iterative positioning scheme that propagates Euclidean distance estimates to a land reference via neighboring nodes. (Figure 3.11). Suppose that NOI *A* has at least two neighbors *B* and *C*, which have already obtained estimates of their Euclidean distance to the anchor node *L*. Further suppose that node *A* has also measured estimates for distances *AB*, *AC*, and that nodes *B* and *C*, which are also assumed to be neighbors, have communicated their distance *BC* to *A*. In this scenario, the lengths of all sides of the quadrilateral *ABCL*, and one of its diagonals *BC*, are known. This allows NOI *A* to compute the second diagonal *AL*, which corresponds to its distance to the land reference. Clearly, this scheme allows node *A* to infer its distance to land reference *L* by means of distance knowledge that was previously acquired by neighboring nodes *B* and *C*, possibly through the same propagation mechanism with the help of other neighboring nodes. In fact, node *A* may now be available to assist other nodes in their calculation of range to that same land reference *L*.

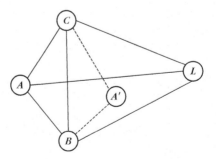

FIGURE 3.11

Euclidean APS scheme.

The previously described propagation mechanism may be applied by every node in the network to find their range to $N \geqslant 3$ land references. Once this is done, one of the multilateration techniques as described in Section 3.1 may be used to obtain final node position estimates.

Note in Figure 3.11 that there is a possibility that node A lies to the right of the BC line; this possibility has been depicted here using node A'. In this case, the distance to node L will be different. The choice between the two possibilities should be made locally by node A, either by comparison to other neighbors that already have an estimate of range to L, or by examining the relation with other common neighbors of B and C. Node A may have to delay its decision until enough neighboring nodes with estimated Euclidean distances to L are available to render the comparisons reliable.

Distance Vector–Bearing APS

DV-bearing is similar to the Euclidean method in the sense that bearing information available at neighboring nodes is propagated iteratively to other nodes that have no AOA information available to estimate their positions using a multilateration scheme. Referring to Figure 3.12, suppose that NOI A knows its bearings to immediate neighbors B and C. These bearings have been denoted by angles $\hat{b}$ and $\hat{c}$, respectively. Note that the angle measurement is done with respect to a reference axis established arbitrarily by node A. Nodes B and C, in turn, know their bearings to a distant land reference node L. The problem is for A to find its bearing to L (denoted by the dashed arrow in Figure 3.12). If B and C are neighbors, then A has the possibility to find all the angles in triangles ABC and BCL. This would allow A to find the angle $\widehat{LAC}$, which yields the bearing of A with respect to L as $\hat{c} + \widehat{LAC}$. In this sense, neighbors B and C have propagated their bearing knowledge, allowing A to find its bearing to L. Now A may

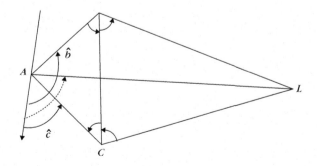

FIGURE 3.12

DV-bearing scheme.

assist other closest neighbors in acquiring their bearings to L as well. In the end, when every node in the network has been able to estimate at least two bearings with respect to two land reference nodes, AOA-based multilateration techniques such as those treated in Section 3.1 may be used to obtain their final position estimates.

Distance Vector–Hop APS

DV-hop is a very simple localized multihop localization scheme that translates the number of hops in a multihop link into distance in meters by means of a correcting factor that is propagated through the network by the anchor nodes. This method has two stages, calibration and multilateration.

Suppose that N anchor nodes are available and that these nodes have perfect knowledge of their own position and of the position of all other anchor nodes in the network such that they may calculate all the true Euclidean distances between each other. Then, if the i-th and j-th anchor nodes establish a communication path via a multihop link, they will be able to relate the number of hops M_{ij} in the route to the actual distance d_{ij} between them to find a correcting factor that corresponds to the number of meters per hop as

$$\text{CHOP}_{ij} = \frac{d_{ij}}{M_{ij}} = \frac{\sqrt{(x_i - x_j)^2 + (y_i - y_j)^2}}{M_{ij}}. \tag{3.40}$$

If every anchor node in the network establishes a multihop link with all other anchor nodes (or a subset of them), an overall average correction factor may be obtained and broadcast to all the nodes in the network to complete the calibration stage.

Now consider an NOI that establishes communication with an anchor node through an M-hop multihop link. The NOI will then be able to estimate its range to the anchor node by simply multiplying the latest broadcasted average correcting factor by the number of hops in the route. Finally, if the NOI is able to establish multihop communication links with $N \geq 3$ anchor nodes, then a multilateration problem may be solved using one of the range-based geometric or statistical multilateration methods treated in Section 3.1.

DV-hop has the advantage of being quite simple and robust to distance estimation errors since it does not depend on TOA, TDOA, or RSS measurements. On the other hand, as has been shown in He et al. [6] and Perez [16], performance of the algorithm degrades in low node-density scenarios and in the presence of nonuniform node distributed networks. These situations will cause nonuniform hop sizes at each multihop route and a nonhomogeneous number of meters per hop at various network regions. The accuracy of the correcting factor calculation also degrades as the reachability radii of the network nodes increases, since fewer hops are required to complete a link, and thus fewer observations are available to accurately infer the hop size in meters. Another disadvantage of this algorithm is the increase in communication overhead caused by the anchor node–to–anchor node connections required to calculate correcting factors, and by the necessary broadcast of updated correcting factors to all nodes in the network.

Although DV-hop can be seen as a localized scheme where an NOI estimates its position via the multihop connection to N land references, this algorithm may also be used as a nonlocalized scheme where every node in the network estimates its position [6]. In this setting, one anchor node broadcasts a beacon to be flooded throughout the network containing the anchor's position with a hop-count parameter initialized to one. Each receiving node maintains the minimum counter value per anchor of all beacons it receives and ignores those beacons with higher hop-count values. Beacons are flooded outward with hop-count values incremented at every intermediate hop. With this, all nodes in the network, including all other anchors, get the shortest distance, in hops, to every anchor. In the end, all nodes have a table that contains the set of hop-count values obtained from each of the available anchors, the anchor true coordinates, and the average correcting factor corresponding to the average hop length in meters. With this information, every node in the network may convert the hop counts to actual distances and apply a multilateration technique among those described in Section 3.1.

Distance Vector–Distance APS

DV-distance is a variation of the DV-hop scheme in which hop counts are substituted by actual distance estimation between two nodes in an intermediate link. M distance measurements $d_i, i = 1 \ldots M$, are collected in an M-hop multihop link between the i-th and j-th anchor nodes and added together to obtain an estimate of the total range between them such that $\hat{d}_{ij} = \sum_{i=1}^{M} d_i$. If the multihop route does not consist of a straight line connecting the two anchor nodes, then the addition of these M distances will yield an *overestimate* of the true range, that is, $\hat{d}_{ij} \geqslant d_{ij}$. However, since the true range between the anchor nodes is known, a correcting factor may be calculated as follows:

$$\text{CDIST}_{ij} = \frac{\hat{d}_{ij}}{d_{ij}}. \tag{3.41}$$

The multilateration stage of the DV-distance algorithm is exactly the same as that corresponding to DV-hop. An NOI communicates with $N \geqslant 3$ anchor nodes via multihop links and obtains range observations based on the correcting factor. Then it performs multilateration with the resulting range measurements to obtain a final position estimate of the NOI.

DV-distance requires a uniform geographic distribution of nodes, nodes with similar transmission power and antenna gains, and a large node density in order for hop sizes to be fairly constant throughout the network. This method does not take into consideration measurement error statistics, and its performance may degrade considerably in the presence of estimation noise and shadowing since it relies on actual distance estimates that may be obtained through TOA or RSS observations. This algorithm is very sensitive to channel variations due to the fact that the correcting factor estimation becomes less precise in the presence of irregular radio reception patterns. Examples of this behavior are presented in [6].

3.4.4 Dead-Reckoning

Dead-reckoning (DR) is a method that has been historically used for navigation. It consists of the estimation of location based on consecutive distance and direction of travel estimates departing from the last known position or fix [19]. In the multihop scenario, DR consists of a localized iterative localization technique that does not require the NOI to be at a one-hop proximity from at least three anchors. Instead, the requirement is that at least three land references are able to establish a link with any NOI via multiple hops. The method relies solely on multiple independent observations coming from multihop paths originating at distinct land references and ending at the NOI.

DR-based localization is similar to the DV-distance algorithm except that in the former, instead of using distance-correcting factors, true range and AOA estimates are obtained at every hop that forms a path from the land reference to the NOI. From now on we will refer to the multihop path that connects a land reference to an NOI as a dead-reckoning path (DRP). Effectively, under noiseless conditions, one DRP will be enough to estimate the position of an NOI if range and AOA measurements are available at each hop. However, under more realistic scenarios where estimates of range and AOA contain errors, multilateration may be used with as many DRPs as can be established by various land references in the network to reduce the effects of range and AOA estimation inaccuracies as well as of error accumulation that arises from the vectorial sum of noisy range and AOA pairs.

The advantages of DR-based multihop localization schemes are that they are robust to nonuniform node positions and unequal node transmission powers and antenna gains, do not require very large node densities to obtain accurate results, avoid the need for the land references to be in constant communication with each other in order to estimate correction factors and the need to broadcast these factors continuously to the entire network. These last points allow a considerable reduction in communication overhead and power consumption in the network.

Analogous to the DR navigational method, location of an NOI may be estimated based on the vectorial sum of range and AOA pairs measured at each of the various multiple hops that link a fixed land reference to that node. Figure 3.13 shows an M-hop DRP that originates at the land reference and ends at the NOI. Figure 3.14 shows a close-up of the i-th hop in this DRP. At each hop a pair of nodes communicates and estimates their range ρ_i and relative angle α_i (measured with respect to the x-axis), $i = 1, \ldots, M$. If no measurement errors exist, the vectorial sum of range-angle pairs over all hops yields a resultant vector with magnitude d and angle θ that uniquely defines the position of the

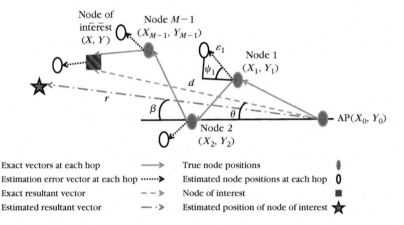

FIGURE 3.13

Dead-reckoning scheme for estimating position of an NOI.

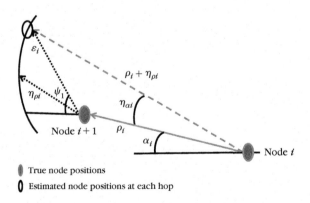

FIGURE 3.14

Scheme for the i-th hop conforming to the DRP.

NOI. Due to range and AOA measurement errors at each hop, denoted by $\eta_{\rho,i}$ and $\eta_{\alpha,i}$, $i = 1, \ldots, M$ in Figure 3.14, the resultant vector will differ from the true position vector by $|d - r|$ meters in magnitude and by $\theta - \beta$ degrees in direction.

Let us envision a set of nodes with TOA and AOA estimation capabilities. AOA estimation schemes are usually considered costly, computationally expensive, and bulky (since they may require large antenna arrays). Note, however, that technology development trends in wireless networks favor the use of higher-frequency bands that will allow the implementation of smaller antenna arrays on the network nodes. Further, recent developments propose accurate and computationally simple AOA estimation schemes that require small antenna arrays with a reduced number of elements [8, 10, 18, 22]. Further, it has been shown in Perez [16] that the mean-square error performance of the DR localization scheme remains fairly insensitive to AOA errors as large as 12 degrees. This means that this algorithm does not require very-high-resolution AOA estimators to perform accurately.

Consider a network with computationally limited nodes where joint TOA/AOA estimation at each hop becomes prohibitive. In this scenario, signal measurements necessary for the estimation of TOA/AOA pairs at each hop may be acquired by the receiving node and propagated through the DRP. Upon the completion of the M-hop DRP, the complete set of signal measurements may be transmitted back to a computationally powerful land reference (possibly through the same DRP route) that may use these data to calculate the M corresponding TOA/AOA pairs. Depending on the TOA/AOA estimation algorithm, large data sets acquired at each hop may be compressed to a sufficient statistic without any loss of estimation accuracy. This will allow a significant reduction of overhead traffic.

DR schemes require a single DRP originating at a single land reference to obtain a node position estimate described by the resultant range-angle pair (r, β). However, if the network environment contains more than three land references, the resulting vector angle β could be ignored and the consequent direction ambiguity avoided by observing the estimated resultant vector magnitudes r_i, $i = 1, \ldots, N$ obtained from DRPs originating at $N \geq 3$ different land references. Then optimal WLS estimates of position could be obtained by multilateration. With this approach the localization scheme profits from multiple independent range observations to improve localization performance via a statistical optimization method, and it allows for the improvement of poor position estimates that may arise from single DRPs conformed by nodes with low-cost, small, and low-complexity antenna array systems with low-resolution AOA estimation capabilities.

Thus, consider the estimation of a specific NOI with unknown coordinates (x, y) using a single DRP between a land reference with known coordinates (x_0, y_0) and that node (see Figures 3.13 and 3.14). The intermediate nodes that link the NOI with the land reference conform to the M hops of the DRP. The coordinates of these intermediate nodes will be denoted as (x_i, y_i), $i = 1, \ldots, M - 1$ and $x_M = x$, $y_M = y$. We define the magnitude of the location vector observation as

$$||\hat{\mathbf{p}}|| = ||\mathbf{p} + \mathbf{e}|| = ||\sum_{i=1}^{M} (\boldsymbol{\gamma}_i + \boldsymbol{v}_i)||, \tag{3.42}$$

where $\mathbf{p}$, $\hat{\mathbf{p}}$, and $\boldsymbol{\gamma}_i$ are vectors described by the magnitude-angle pairs (d, θ), (r, β), and (ρ_i, α_i), respectively, and $\rho_i = [(x_i - x_{i-1})^2 + (y_i - y_{i-1})^2]^{1/2}$, $\alpha_i = \tan^{-1}[(y_i - y_{i-1})/(x_i - x_{i-1})]$, $i = 1, \ldots, M$. Also, vector $\boldsymbol{v}_i$, which is described by the magnitude-angle pair (ε_i, ψ_i), $i = 1, \ldots, M$, corresponds to the measurement error vector at the i-th hop of the DRP (see Figures 3.13 and 3.14).

Following a simple trigonometric calculation, it can be shown that at the i-th hop

$$\varepsilon_i = \{2\rho_i[1 - \cos(\eta_{\alpha,i})][\rho_i + \eta_{\rho,i}] + \eta_{\rho,i}^2\}^{1/2}, \tag{3.43}$$

and

$$\psi_i = \pi + \alpha_i - \cos^{-1}\left[\frac{\rho_i - (\rho_i + \eta_{\rho,i})\cos(\eta_{\alpha,i})}{\varepsilon_i}\right]. \tag{3.44}$$

Hence, the error vector at the i-th hop, $\boldsymbol{v}_i$, depends on the ranging and AOA estimation errors $\eta_{\rho,i}$ and $\eta_{\alpha,i}$, and on the range ρ_i and angle α_i between the nodes conforming to the i-th hop.

In this treatment we may assume that a routing algorithm exists that chooses the shortest path between a land reference and the NOI congruent with two basic limitations—node density D and node coverage range R. It may also be assumed that nodes remain quasistatic in the time interval spanning the propagation and processing delay at each hop so that a DRP, with corresponding TOA/AOA estimation pairs, is established in a scenario where the positions of the NOI and intermediate nodes remain fairly constant during the multihop routing process. The quasistatic node assumption is reasonable, for example, in mobile portable computing network scenarios where nodes (users) tend to stay at a fixed position for long periods of time.

Routing angle, α_i, may be modeled as a random variable symmetrically distributed around θ degrees with standard deviation σ_{spread} also denoted as routing angular spread. Parameter σ_{spread} describes the average angular amount by which each hop on the actual routing sequence deviates from being collinear to the direct resultant range vector defined by the pair (d, θ) as shown in

Figure 3.13. Its value will be directly determined by the network's node density D and the node coverage range R.

Ranging estimation errors $\eta_{\rho,i}$ $(i = 1, \ldots, M)$ that arise from TOA, TDOA, or RSS measurement inaccuracies may be modeled as independent zero mean random variables with variance $\sigma_{\eta_\rho}^2$. In a similar manner, angular errors $\eta_{\alpha,i}$ $(i = 1, \ldots, M)$ will depend on the optimality of the AOA estimators in the presence of additive noise and other channel impairments. These errors may also be assumed independent, with zero mean and variance $\sigma_{\eta_\alpha}^2$. In the end, the overall vectorial sum for a DRP will yield a resulting vector whose magnitude $\|\hat{\mathbf{p}}\|$ will consist of the sum of the true distance between the land reference and the NOI, and an error term as given in Equation (3.15) with functions $f_i(x, y)$ corresponding to Euclidean distances between the land references and the NOI.

Consider $N \geqslant 3$ DRP position vector estimates. If we ignore the estimated resultant vector angle β at each of the N DRPs, the position of an NOI may be estimated by applying a multilateration algorithm based on N observations that contain true distance magnitudes corrupted by additive noise.

Define a vector containing the true coordinates of the NOI as $\mathbf{q} = [x \, y]^H$, and let $N \geqslant 3$ land references, with coordinates $\{x_{LR,i}, y_{LR,i}\}, i = 1, \ldots N$, be available in the network, and the i-th land reference establish a link to the NOI via M_i hops. The N resultant range observations are collected in a vector to obtain

$$\mathbf{r} = \mathbf{d}(\mathbf{q}) + \boldsymbol{\eta}, \tag{3.45}$$

where $\mathbf{r} = [r_1 \, r_2 \ldots r_N]^H$, $\mathbf{d}(\mathbf{q}) = [d_1(\mathbf{q}) \, d_2(\mathbf{q}) \ldots d_N(\mathbf{q})]^H$, $\boldsymbol{\eta} = [\eta_1 \, \eta_2 \ldots \eta_N]^H$, and $d_i(\mathbf{q}) = [(x_{LR,i} - x)^2 + (y_{LR,i} - y)^2]^{\frac{1}{2}}$ is the true Euclidean distance between the unknown node position and the i-th land reference. Noise samples η_i and η_j $(i \neq j)$ may be assumed independent; however, they are not identically distributed unless all DRPs have the same number of hops. The independence assumption will be reasonable whenever the node density and reachability radius are large enough, and the measured resultant distances come from DRPs originating at sufficiently separated land references such that no route shares common hops.

The last step in the DR algorithm consists of applying the LS or WLS multilateration process, described in Section 3.3.1, to the N-dimensional vector of noisy distances given in Equation (3.45).

3.5 PERFORMANCE ASSESSMENT OF LOCATION ESTIMATION SYSTEMS

In this section, we present some performance measures that describe the accuracy of a location estimation scheme based on the probability distribution of range- or AOA-related observations.

3.5.1 Cramer-Rao Bound

The problem that has been treated in this chapter is that of estimating the NOI coordinates vector $\mathbf{q}$ given a set of noisy observations collected in vector $\mathbf{r}$ as expressed in Equation (3.15).

The Cramer Rao bound (CRB) provides a theoretical lower limit to the error covariance matrix $\mathbf{C}$ of any unbiased estimator of coordinate vector $\mathbf{q}$ such that

$$\mathbf{C} = E\{[\hat{\mathbf{q}} - \mathbf{q}][\hat{\mathbf{q}} - \mathbf{q}]^{H}\} \geqslant \mathbf{J}^{-1}, \tag{3.46}$$

where $\mathbf{J}$ corresponds to the Fisher information matrix given by

$$\mathbf{J} = -E\left\{ \frac{\partial}{\partial \boldsymbol{\theta}} \left(\frac{\partial}{\partial \boldsymbol{\theta}} \ln p_{\mathbf{q}}(\mathbf{r}) \right)^{H} \right\}. \tag{3.47}$$

Here, $p_{\mathbf{q}}(\mathbf{r})$ corresponds to the joint probability density distribution of observation vector $\mathbf{r}$.

If the observation noise in Equation (3.15) is Gaussian with zero mean and covariance matrix $\mathbf{K}$, then the observation vector $\mathbf{r}$ is also Gaussian with mean vector equal to $\mathbf{f}(\mathbf{q})$ and with the same covariance matrix $\mathbf{K}$. In this case, the CRB for position estimate vector $\hat{\mathbf{q}}$ is easily calculated as in Scharf [23]:

$$\text{CRB} = \mathbf{J}^{-1} = \frac{\partial \mathbf{f}(\mathbf{q})^{H}}{\partial \mathbf{q}} \mathbf{K}^{-1} \frac{\partial \mathbf{f}(\mathbf{q})}{\partial \mathbf{q}}. \tag{3.48}$$

3.5.2 Circular Error Probability

The circular error probability (CEP) [25] is a simple measure of accuracy defined as the radius of the circle that has its center at the mean and contains half the realizations of a random vector of coordinate estimates. It is a measure of uncertainty in the location estimator $\hat{\mathbf{q}}$ relative to its mean $E\{\hat{\mathbf{q}}\}$. If the location estimator is unbiased, the CEP is a measure of the estimator uncertainty relative to the true NOI position. If the magnitude of the bias vector is bounded by B, then with a probability of one-half, a particular estimate is within a distance of $B + CEP$ from the true position. This concept is illustrated in Figure 3.15.

From its definition, the CEP may be derived by solving the following equation:

$$\frac{1}{2} = \int \int_{R} p_{\hat{\mathbf{q}}}(\boldsymbol{\zeta}) d\zeta_1 d\zeta_2, \tag{3.49}$$

where $p_{\hat{\mathbf{q}}}(\boldsymbol{\zeta})$ is the probability density function of vector estimate $\hat{\mathbf{q}}$, and the integration region is defined as $R = \{\boldsymbol{\zeta} : |\boldsymbol{\zeta} - E\{\hat{\mathbf{q}}\}|\} \leqslant CEP$. Most of the time, a closed-form expression of Equation (3.49) is difficult to find and numerical integration must be performed. However, the following approximation, which is accurate to within 10%, is often used [25]:

$$\text{CEP} \approx 0.75\sqrt{E\{(\hat{\mathbf{q}} - E\{\hat{\mathbf{q}}\})^{H}(\hat{\mathbf{q}} - E\{\hat{\mathbf{q}}\})\}} = 0.75\sqrt{\lambda_1 + \lambda_2} = 0.75\sqrt{\sigma_1^2 + \sigma_2^2}. \tag{3.50}$$

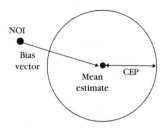

FIGURE 3.15

Geometry of the CEP definition.

Here λ_1 and λ_2 are the eigenvalues of the estimator covariance matrix, which is given by

$$E\{(\hat{\mathbf{q}} - E\{\hat{\mathbf{q}}\})^H (\hat{\mathbf{q}} - E\{\hat{\mathbf{q}}\})\} = \begin{bmatrix} \sigma_1^2 & \sigma_{12} \\ \sigma_{12} & \sigma_1^2 \end{bmatrix}. \tag{3.51}$$

3.5.3 Geometric Dilution of Precision

The geometric dilution of precision (GDOP) [24, 25] provides a measure of the effect of the geometric configuration of the land references on the location estimate. It is defined as the ratio of the root mean-square position error to the root mean-square ranging error. Then, for an unbiased estimator, the GDOP is given as

$$\text{GDOP} = \frac{1}{\sigma_r} \sqrt{E\{(\hat{\mathbf{q}} - E\{\hat{\mathbf{q}}\})^H (\hat{\mathbf{q}} - E\{\hat{\mathbf{q}}\})\}}, \tag{3.52}$$

where σ_r may be seen as the fundamental ranging or AOA estimation error.

The GDOP indicates how much the fundamental ranging error is magnified by the geometric relation among the NOI and the land references. Comparing Equations (3.50) and (3.52), we observe that

$$\text{CEP} \approx (0.75\sigma_r)\text{GDOP}. \tag{3.53}$$

REFERENCES

[1] N. Bulusu, J. Heidemann, D. Estrin, GPS-less low cost outdoor localization for very small devices, IEEE Personal Communications Magazine, 7 (5) (2000) 28–34.

[2] T.F. Coleman, Y. Li, An interior, trust region approach for nonlinear minimization subject to bounds, SIAM Journal on Optimization, 6 (1996) 418–445.

[3] L. Doherty, K.S.J. Pister, L. El Ghaoui, Convex position estimation in wireless sensor networks, in: Proceedings of INFOCOM, 3 (2001) 1655-1663.

[4] W.H. Foy, Position-location solutions by Taylor-series estimation, IEEE Transactions on Aerospace and Electronic Systems, 12 (2) (1976) 187-194.

[5] G. Zhou, T. He, S. Krishnamurthy, J.A. Stankovic, Impact of radio irregularity on wireless sensor networks, in: Proceedings of the 2nd International Conference on Mobile Systems, Applications, and Services, June (2004) 125-138.

[6] T. He, C. Huang, B. Blum, J. Stankovic, T. Abdelzaher, Range-free localization schemes in large scale sensor networks, in: Proceedings of the 9th International Conference on Mobile Computing and Networking, September (2003) 81-95.

[7] O. Hernandez, F. Bouchereau, D. Muñoz, Maximum likelihood position estimation in ad-hoc networks using a dead reckoning approach, IEEE Transactions on Wireless Communications, 7 (5) Part I (2008) 1572-1584.

[8] J. Yli-Hietanen, K. Kalliojarvi, J. Astola, Low-complexity angle of arrival estimation of wideband signals using small arrays, in: Proceedings of the 8th IEEE Signal Processing Workshop on Statistical Signal and Array Processing, June (1996) 109-112.

[9] L. Kovavisaruch, K.C. Ho, Modified Taylor-series method for source and receiver localization using TDOA measurements with erroneous receiver positions, in: Proceedings of the IEEE International Symposium on Circuits and Systems, 3 (2005) 2295-2298.

[10] W. Menzel, A. Gronau, V. Maiyappan, W. Mayer, Small-aperture, high-resolution beam-scanning antenna array using nonlinear signal processing, European Conference on Wireless Technology, October (2005) 435-438.

[11] D. Niculescu, B. Nath, Ad hoc positioning system (APS), IEEE Global Telecommunications Conference, 5 (2001) 2926-2931.

[12] D. Niculescu, B. Nath, Ad hoc positioning system (APS) using AOA, in: Proceedings IEEE INFOCOM, 3 (2003) 1734-1743.

[13] D. Niculescu, Positioning in ad hoc sensor networks, IEEE Network, 18 (4) (2004) 24-29.

[14] N. Patwari, A.O. Hero III, M. Perkins, N.S. Correal, R.J. O'Dea, Relative location estimation in wireless sensor networks, IEEE Transactions on Signal Processing, 51 (8) (2003) 2137-2148.

[15] N. Partwari, J.N. Ash, S. Kyperountas, A.O. Hero III, R.L. Moses, S.D. Correal, Locating the nodes: cooperative localization in wireless sensor networks, IEEE Signal Processing Magazine, 22 (4) (2005) 54-69.

[16] V. Perez, Position estimation using dead reckoning and other localized algorithms in wireless networks, M.S. Thesis Dissertation, August 2008, ITESM, Monterrey, Mexico.

[17] R.A. Poisel, Electronic Warfare Target Location Methods, Artech House, 2005.

[18] Z. Popovic, C. Walsh, P. Matyas, C. Dietlein, D.Z. Anderson, High-resolution small-aperture angle of arrival detection using nonlinear analog processing, IEEE International Microwave Symposium Digest, 3 (2004) 1749-1752.

[19] C. Randell, C. Djiallis, H. Muller, Personal position measurement using dead reckoning, in: Proceedings of the 7th IEEE International Symposium on Wearable Computers, October (2003) 166-173.

[20] A. Savvides, C.C. Han, M.B. Srivastava, Dynamic fine-grained localization in ad-hoc networks of sensors, in: Proceedings of the 7th Annual International Conference on Mobile Computing and Networking, July (2001) 166-179.

[21] A. Savvides, H. Park, M.B. Srivastava, The n-hop multilateration primitive for node localization problems, Mobile Networks and Applications, 8 (4) (2003) 443-451.

[22] H.G. Schantz, Smart antennas for spatial RAKE UWB systems. IEEE Antennas and Propagation Society International Symposium, 3 (2004) 2524-2527.

[23] L.L Scharf, Statistical Signal Processing—Detection, Estimation, and Time Series Analysis, 2nd edition, Addison-Wesley, 1991.

[24] M.A. Spirito, On the accuracy of cellular mobile station location estimation, IEEE Transactions on Vehicular Technology, 50 (3) (2001) 674-685.

[25] D.J. Torrieri, Statistical theory of passive location systems, IEEE Transactions on Aerospace and Electronic Systems, AES-20 (2) (1984) 183-198.

Heuristic Approaches to the Position Location Problem

4

In many disciplines, rigorous solution of a problem is not feasible because of its complexity due to a large number of variables, inadequate knowledge of how the variables interact, long computation times, and high noise environments that mask system functionality, among other factors. Nevertheless, accumulated experience, solutions to similar problems in other disciplines, solution of simplified scenarios, and intuition enable engineers to develop solutions for these kinds of problems. Such methodologies are referred to as *heuristic techniques*.

Heuristic thinking is combined with appropriately applied mathematical tools and simulation to cope with parts of the problem. In this chapter, we consider heuristics applicable to the position location (PL) problem.

Heuristic solutions involve uncertainty that is inversely proportional to the available information. When PL information is insufficient, closed mathematical solutions are not possible. A rough estimation of the location may nonetheless enable applications in which accurate location is not a requirement, as in the case of location-based services where the region or area of interest allows sales without precise knowledge of subscriber location.

In this chapter, heuristics applicable to both single- and multihop scenarios are introduced and expected performance is addressed.

4.1 SINGLE-HOP AND RELATIONAL SCENARIOS

The basic structure of telecommunication systems has enabled one to obtain the location of a subscriber using the information associated with the last mile segment. For instance, in POTS (plain old telephone service) operators were able to say where a call originated based on the record of service installation. Resolution then was limited to customer premises. This approach has been shown to be inapplicable to cellular systems where subscribers are meant to have unrestricted mobility. In the basic cellular architecture, a control center connects the base stations that are meant to provide service in designated areas. Thus, subscriber location can be determined with greater or lesser uncertainty, depending on size of service area.

Location uncertainty can be given by location entropy:

$$H(L) = - \int_{l \in L} f_L(l) \log f_L(l) dl, \tag{4.1}$$

where $f_L(l)$ denotes the probability density function of a unit being located at the l site in the location domain L. Location domain will depend on cell site. Thus, uncertainty tends to be great for macro-cells and small for micro- and pico-cells. In the absence of further information, $f_L(l)$ is assumed to be uniformly distributed in the location domain. Nevertheless, knowledge of attraction poles within a coverage area as well as traveling patterns contributes to reducing location entropy. Entropy deals with average location uncertainty rather than location realization for a given individual.

To provide continuity of service during mobile journeys, cell coverage areas are designed not as partitions but as overlapping areas that enable graceful handoff algorithms. Although handoff is not a concern in this chapter, it is known that subscribers in the cell-to-cell transition area are able to communicate with or listen to several base stations. This information enables one to establish not only the base stations a subscriber is connected through but also the subscriber's likely geographic neighborhood.

Eventually a subscriber location can be established with a very small uncertainty (e.g., when the mobile is reachable from a pico-cell). Nevertheless, as time elapses, the uncertainty grows. This can be described using diffusion process modeling where the $f_L(l)$ variance tends to grow with time and the mean location can be calculated if subscriber mobility trends are known. Such mobility patterns can be computed for a population or an individual, with higher costs associated with finer resolution. For instance, a hypothetical individual in a mall could be traced from his electronic transactions or the hot spots he passes. When he leaves the mall area, his whereabouts can be computed based on the cell coverage areas that he passed through.

Location determination precision is always dependent on the accuracy and availability of the parameters used in the location acquisition process. In cellular systems, subscribers are reachable in a single hop from one or several base stations with well-known locations. In the case of ad hoc networks, information travels along multiple nodes of imprecise location so that uncertainty grows with the number of links.

Nevertheless, as a packet progress along a path, some information, concerning the nodes the packet has passed through, is gathered. In this form it is possible to know the number of hops a node is from a given access point (AP). Some additional information may also be available; for instance, a node can inform the other nodes that it can listen to or that are within its reachability range. Further information may contain distance-dependent measurements such as field strength and, in more complex systems, time delay or angular observations. These allow dead-reckoning techniques that were considered in Chapter 3 from the perspective of estimation theory.

An important assumption in ad hoc and sensor networks is that subscribers relay information to/from nearby neighboring nodes so that messages propagate from site to site until a sink node or network gateway is reached. These sink nodes, also called access points (APs), are placed at convenient sites. Thus, APs not only provide connectivity to other networks or systems, but also become geographic references. In this chapter, we consider a connected network in the sense that all nodes are linkable to/from an AP. Otherwise, the system is not even aware of the existence of such a node.

4.1.1 Range-Free Location Estimation Systems

Location determination is only possible in relation to an agreed-on reference. In many cases, the location process invokes multiple observation sites that supply information for centralized information processing, and it is assumed that the node can be reached from observation sites in a single step. These observations imply range-related measurements such as field intensity or delay differences that are followed by trilateration processes that consider intersection of conic forms. These trilateration processes require at least three observation points, although it may not be possible to ensure that propagation conditions are equally adequate to and from all observation points. For instance, if the node is in close proximity to a base station, it tends to be well apart from other observation points. (Good visibility to and from other distant base stations may require higher transmission power that also impacts higher interference levels.)

Technologic advances in antenna technology allow determining angular location and this scheme permits the use of only two observation sites. If coverage of the region cannot be guaranteed, some technologies provide for the introduction of supplementary location mobile units (LMUs) [1, 2] (see Figure 4.1a). Assuming that the base station/AP and an LMU are located at coordinates $(-D, 0)$ and $(D, 0)$, they will provide for the node at $P(x, y)$ angular observations

$$\phi_1 = tg^{-1}\left(\frac{y}{D+x}\right)$$

and

$$\phi_2 = tg^{-1}\left(\frac{y}{D-x}\right);$$

the node is assumed to be at the intersection of lines $y = [tg\,\phi_1](x+D)$ and $y = [tg\,\phi_2](D-x)$.

Location will be uncertain when the node is on the line connecting the AP and the LMU. Precise angular observations are prevented by multipath phenomena and the location error sensitivity to angular errors will strongly depend on the (x, y) location. Figure 4.1b shows mean square error compared to angular error. Gaussian error distribution has been assumed.

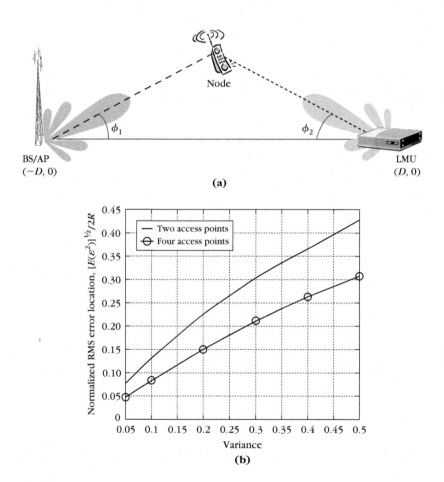

FIGURE 4.1

Angular range-free location. (a) Angular location scenario. (b) Error performance versus angular error variance.

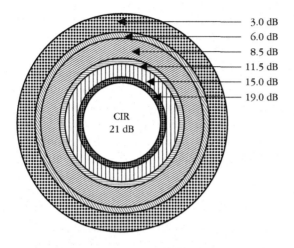

3.0 dB
6.0 dB
8.5 dB
11.5 dB
15.0 dB
19.0 dB

CIR
21 dB

FIGURE 4.2

CIR location regions.

Other range schemes may relate location information to quality-of-service figures. For instance, systems like WiMax or LTE are able to dynamically assign throughputs depending on the perceived carrier-to-interference ratio (CIR). That is, quality-of-service parameters may be associated with location areas of subscriber units as illustrated in Figure 4.2. In general, since CIRs tend to monotonically decrease as separation distance between subscriber unit and the base station increases, we can say that for a CIR$>\zeta_i$, $\mathbf{n}_0$ will belong to the region

$$R_i = \{(x, y) \,\Big|\, \sqrt{x^2 + y^2} \leqslant \rho_i\}$$

and for $\zeta_i < \text{CIR} \leqslant \zeta_{i+1}$, $\mathbf{n}_0$ will be said to be in the ring $R_i \cap R^c_{i+1}$. If quality parameters associated with different base stations are available, the subscriber unit is said to be in the intersection of the corresponding rings.

4.1.2 Signal Signature

Individuals in a group can be identified by the collection of characteristics that distinguish each individual. For instance, in voice recognition, format location and its relative strengths can be used for individual classification.

In wireless systems, a user receives signals from various radio sources, and in the ideal scenario the associated characteristic for each signal can be used for location estimation.

As an example, let us consider a scenario where the mobile can see/be seen from several base stations BS_i $i = 1, \ldots, k$. Each location $\mathbf{n}_0$ will have an associated vector or location signature (e.g., $\mathbf{F}_o = (F_{o1}, \ldots, F_{ok})$) where F_{oi} may denote the field intensity associated with BS_i. Once subscriber observations are obtained, they will be compared against the reference signatures and the closest reference will be used to estimate subscriber location.

In practice, an ideal mapping $\mathbf{n}_0 \leftrightarrow \mathbf{F}_o = (F_{o1}, \ldots, F_{ok})$ may depart heavily from practical observations due to clutter and other propagation impairments. Therefore, the signature space will be adapted on the basis of multiple observations. And a biunique mapping cannot always be guaranteed. In addition to the above propagation noise effects, mobiles may estimate parameters with low resolution, producing further coarseness. This is illustrated in Figure 4.3 where a linear location d is mapped into a 2D signal-strength signature [3].

Although the described methodology resembles some of the issues in estimation theory, no statistical channel characterization is required.

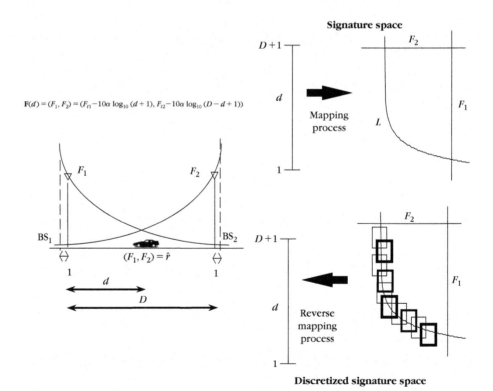

$$\mathbf{F}(d) = (F_1, F_2) = (F_{r1} - 10\alpha \log_{10}(d+1), F_{r2} - 10\alpha \log_{10}(D-d+1))$$

FIGURE 4.3

Signal-strength signature scheme.

4.2 MULTIHOP SCENARIOS

In cellular systems, subscribers are reached via links connecting a well-known location base station to the subscriber unit. During the handoff process or for location purposes, several single-hop links are established to various base stations, enabling a triangulation process suitable for location determination. Current technological trends favor multihop scenarios in order to save transmission power or for better coverage planning. In single-hop scenarios, the accuracy of the location process will be dependent on the measurement error of the involved parameters.

In multihop scenarios, as in mesh and ad hoc networks, nodes will forward the information to the next node in the path, and this will relay the information to a more distant node. This process continues until the destination node is reached. Since the location of the intermediate nodes is not known, location uncertainty increases with the number of relay stages.

Position location in the absence of direct landmarks demands heuristic approaches that provide a picture of the likely location. In the following sections, several heuristics applicable to multihop scenarios are described.

4.2.1 Triangle Concatenation

Triangles are rigid bodies that are determined by three parameters. For instance, given one side, knowledge of the two other sides or two angles or a combination of these will be sufficient for triangle specification.

In ad hoc networks, each time a packet goes through a node, a label may be added to the packet so that all the existing links are known at the control center. This information can be fed into an $n \times n$ square array (adjacency matrix $\mathbf{A}_{n \times n} = \{a_{ij}\}$), where n is the number of nodes (including the APs) and a_{ij} is set to 1 if a link connecting nodes $\mathbf{n}_i$ and $\mathbf{n}_j$ exists; $a_{ij} = 0$ otherwise. A self-adjacent coefficient is defined as $a_{jj} = 0$.

For definition convenience and since links are assumed to be bidirectional, $\mathbf{A}$ is a symmetric matrix. The table in Figure 4.4a exemplifies the adjacency matrix for the network in the Figure 4.4b. We define the node $\mathbf{n}_i$ connectivity index c_i as the product $c_i = [a_{i1}, \ldots a_{in}] \times [1, \ldots \ldots, 1]^T$. Note that the values $c_i = 0$ do not occur, as the network is connected (i.e., there is a path connecting any two nodes in the network). $c_i = 1$ corresponds to nodes with a single connection. If $c_i \geqslant 2$, we can search for triangles that include $\mathbf{n}_i$ as a vertex. In brief, find all the k values such that $a_{ik} = 1$; if $a_{ik} = 1, a_{ij} = 1$, and $a_{jk} = a_{kj} = 1$, we say that the triangle $T(\mathbf{n}_i, \mathbf{n}_j, \mathbf{n}_k)$ exists. Triangles with two nodes in common, such as triangles $T(\mathbf{n}_{12}, \mathbf{n}_{10}, \mathbf{n}_{11})$ and $T(\mathbf{n}_{11}, \mathbf{n}_{10}, \mathbf{n}_9)$ in Figures 4.4a and b, are said to be concatenated. All triangle concatenations are completed; note that neither

i \ j	1	2	3	4	5	6	7	8	9	10	11	12
1	0	1	1	0	0	0	0	0	0	0	0	0
2	1	0	1	1	0	0	0	0	0	0	0	0
3	1	1	0	1	1	0	0	1	0	0	0	0
4	0	1	1	0	1	0	0	0	0	0	0	0
5	0	0	1	1	0	0	0	1	0	1	0	0
6	0	0	0	0	0	0	1	0	0	0	0	0
7	0	0	0	0	0	1	0	1	1	0	0	0
8	0	0	1	0	0	0	1	0	1	1	0	0
9	0	0	0	0	0	0	1	1	0	1	1	0
10	0	0	0	0	1	0	0	1	1	0	1	1
11	0	0	0	0	0	0	0	0	1	1	0	1
12	0	0	0	0	0	0	0	0	0	1	1	0

(a)

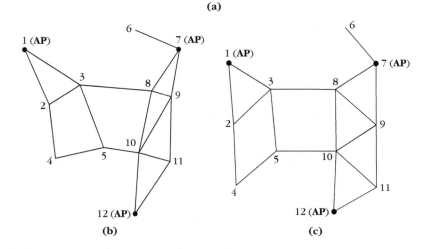

(b)   (c)

FIGURE 4.4

Triangle concatenation scheme. (a) Adjacency matrix $\mathbf{A}_{n \times n} = \{a_{ij}\}$. (b) Actual network. (c) Obtained network.

triangle nor concatenation can exist in isolation, but each will connect to an AP. As a matter of fact, two APs can be vertices of a triangle concatenation and the concatenation can be shifted, rotated, and scaled to match the actual APs' locations. Thus, an approximate picture of the network is provided. Note that no distance measurements were available. Consequently, the constructed triangles were assumed equilateral (e.g., see concatenation $T(\mathbf{n}_{12}, \mathbf{n}_{10}, \mathbf{n}_{11})$, $T(\mathbf{n}_{11}, \mathbf{n}_{10}, \mathbf{n}_9)$, $T(\mathbf{n}_8, \mathbf{n}_{10}, \mathbf{n}_9)$, and $T(\mathbf{n}_8, \mathbf{n}_7, \mathbf{n}_9)$ in Figure 4.4c, where nodes $\mathbf{n}_{12}$ and $\mathbf{n}_7$ are meant to be APs).

There will always be ambiguities. For instance, triangle $T(\mathbf{n}_1, \mathbf{n}_2, \mathbf{n}_3)$ contains an AP but it is not part of a triangle concatenation. Nevertheless, the length of the edges was set using the scale of another concatenation. In order to keep geometric scenarios simple, edge crossings are assumed to be nonexistent, so vertices $\mathbf{n}_4$, $\mathbf{n}_2$, $\mathbf{n}_3$, and $\mathbf{n}_5$ define a quadrilateral shape and the side length will be defined according to the edges in $T(\mathbf{n}_1, \mathbf{n}_2, \mathbf{n}_3)$. Edges $(\mathbf{n}_3, \mathbf{n}_8)$ and $(\mathbf{n}_5, \mathbf{n}_{10})$ tend to be commensurate with the already-defined edges. Node $\mathbf{n}_6$ belongs to only one edge and can be set, considering that it is unlikely to be in close proximity to $\mathbf{n}_8$; otherwise, an edge might exist and edge $(\mathbf{n}_7, \mathbf{n}_6)$ length is also meant to be commensurate to the adjacent edge $(\mathbf{n}_7, \mathbf{n}_8)$. (Two edges are said to be adjacent when they have a node in common.)

It will be impractical to cover all eventual possibilities and ambiguities. In any case, this approach provides only a rough idea of node distribution. Improved strategies will be described in subsequent sections.

4.2.2 Random Flight

Location determination will always involve connection to well-known landmarks. Therefore, the PL process will be dependent on the way that the node-to-AP paths are established. Each link in a path can be treated as a vector with a given magnitude and direction, and the overall route will be a concatenation of random links in the path.

Random vector addition has many applications and has been studied from multiple perspectives and applications, including Brownian motion, which led to diffusion models, polymer modeling, and so on [4, 5]. These models often include limited or no restriction considerations, while in ad hoc networks, the random paths exhibit some orientation determined by the location of the end nodes. Perhaps the most similar scenario to an ad hoc one is the dead-reckoning process.

In early navigation, dead-reckoning considered a vessel that set its traveling direction aimed at a destination site. Tides and wind move the vessel away from the intended direction. Thus, after a given time, travel distance and direction are estimated and the vessel is reoriented.

In the ad hoc environment, the travel distance or hop length as well as the direction will depend on location nodes. Routing schemes are not considered in this chapter, although it is assumed that at each stage a hop in a direction close to the destination is selected. In this form, the vector linking $\mathbf{AP}_A$ to $\mathbf{n}_0$ is the sum of the individual hop vectors in the path

$$(\mathbf{AP}_A \rightarrow P_{AN_1}(1) \rightarrow \cdots \rightarrow P_{AN_j}(k) \rightarrow \cdots \rightarrow \mathbf{n}_0).$$

In dead-reckoning, a fix is periodically conducted so that the correction prevents error accumulation. Because fixes may not be possible in ad hoc networks, errors tend to grow with the number of hops [6]. Dead-reckoning has been analyzed in Chapter 3 from the perspective of estimation theory. However, in some cases, knowledge of whole-hop vectors (magnitude and angle) may not be available. But range estimates can be used for location estimation, as we describe in the next section.

Pyramidal Approach

Some of the previous procedures assume that hop distances are not available, but that the average hop length is. In this section, it is assumed that noisy range observations are available, although no angular measurements are assumed.

Let us consider a network area defined by three, $\mathbf{AP}_A$, $\mathbf{AP}_B$, and $\mathbf{AP}_C$, with known locations (x_A, y_A), (x_B, y_B), (x_C, y_C), respectively (see Figure 4.5). Let $\mathbf{n}_0$ be a node whose unknown location (x_0, y_0) will be estimated. If actual distances between node $\mathbf{n}_0$ and the APs were available, the location (x_0, y_0) could be unequivocally determined by triangulation. However, this process cannot be directly invoked, as connection of node $\mathbf{n}_0$ to $\mathbf{AP}_A$, is achieved after the concatenation of multiple hops (say,

$$\mathbf{AP}_A \leftrightarrow P_{AN_1}(1) \leftrightarrow \cdots \leftrightarrow P_{AN_j}(k) \leftrightarrow \cdots \leftrightarrow \mathbf{n}_0,$$

where $P_{AN_0}(k)$ denotes the k-th successive intermediate node in the path linking $\mathbf{AP}_A$ to $\mathbf{n}_0$). In principle, all node locations in the path are unknown but $\mathbf{AP}_A$. Nevertheless, Euclidean distance $d\langle \mathbf{AP}_A, \mathbf{n}_0 \rangle$ between nodes at the k-th link may be estimated using, for instance, time of arrival (TOA) measurements [7, 8]. A first rough estimate δ_A of the distance between node $\mathbf{n}_0$ and $\mathbf{AP}_A$ can be obtained by adding the hop lengths of the path:

$$\delta_A = d\langle \mathbf{AP}_A, P_{A1} \rangle + \Delta_1 + d\langle P_{A1}, P_{Ak} \rangle + \Delta_k + \cdots,$$

$$\cdots + d\langle P_{Az-1}, \mathbf{n}_0 \rangle + \Delta_z, \tag{4.2}$$

where Δ_k stands for the measuring length error in the k-th hop, which we assume to be exponentially distributed. Note that even if the terms Δ_k are neglected, according to the triangle inequality, $d\langle \mathbf{AP}_A, \mathbf{n}_0 \rangle \leqslant \delta_A$. So δ_A is an overestimate of the true distance $d\langle \mathbf{AP}_A, \mathbf{n}_0 \rangle$ between $\mathbf{n}_0$ and $\mathbf{AP}_A$. Equality of δ_A to the actual distance will be achieved only when all the nodes in the link are collinear to the straight line passing through $\mathbf{n}_0$ and $\mathbf{AP}_A$, and provided that measurements $d_{Aj-1,Aj} = d(P_{Aj-1}, P_{Aj})$ are noiseless (i.e., $\Delta_j = 0$).

In a similar manner, overestimates δ_B and δ_C corresponding to distances from $\mathbf{n}_0$ to $\mathbf{AP}_B$ and $\mathbf{AP}_C$, respectively, can be obtained. Note that distance overestimates δ_A, δ_B, and δ_C may induce errors in different directions. However, these errors may tend to compensate each other [9]. With this in mind, and in order to maintain low hardware and operational complexity, the location (x_0, y_0) may be heuristically inferred by noting that edges δ_A, δ_B, and δ_C define the vertex V of a pyramid with base $\mathbf{AP}_A$, $\mathbf{AP}_B$, and $\mathbf{AP}_C$ as illustrated in Figure 4.5. This vertex can be orthogonally projected as point $\mathbf{n}_0 = (x, y)$ on the pyramid

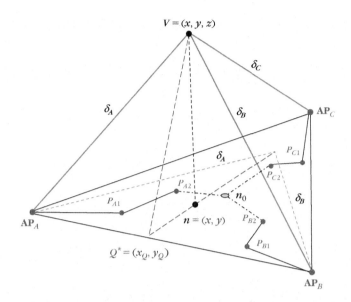

FIGURE 4.5

Pyramidal location.

base and it may be adopted as the PL estimate. Coordinates (x, y) are meant to be in the feasibility region defined by the pyramid base and they can be readily obtained by a simple algebraic analysis in the following way: Given the reference point locations $\mathbf{AP}_A$ and $\mathbf{AP}_B$, distance estimates δ_A and δ_B define a triangle with vertices $\mathbf{AP}_A = (x_A, y_A, 0)$, $\mathbf{AP}_B = (x_B, y_B, 0)$, and $V = (x^*, y^*, z)$.

At this stage, vertex V coordinates (x^*, y^*, z) are unknown; select $Q^* = (x_Q, y_Q, 0)$ as the point on the segment $\mathbf{AP}_A \leftrightarrow \mathbf{AP}_B$, such that segments $\mathbf{AP}_A \leftrightarrow \mathbf{AP}_B$ and $Q^* \leftrightarrow V$ are orthogonal. Let $P^* = (x^*, y^*, 0)$ be the projection of V on the plane x-y; segment $P^* \leftrightarrow Q^*$ will define a line also orthogonal to segment $\mathbf{AP}_A \leftrightarrow \mathbf{AP}_B$ [10].

$$y^* - y_Q = \frac{x_A - x_B}{y_B - y_A}(x^* - x_Q), \tag{4.3}$$

where

$$x_Q = \frac{[x_A + r_{AB} \times x_B]}{1 + r_{AB}}, \quad y_Q = \frac{[y_A + r_{AB} \times y_B]}{1 + r_{AB}}$$

and

$$r_{AB} = \frac{\delta_A^2 - \delta_B^2 + d(\mathbf{AP}_A, \mathbf{AP}_B)^2}{\delta_B^2 - \delta_A^2 + d(\mathbf{AP}_A, \mathbf{AP}_B)^2}.$$

Similarly, locations $\mathbf{AP}_C$, $\mathbf{AP}_B$, and distance estimates δ_C and δ_B allow the projection of V in the x-y plane on the line

$$y^* - y_s = \frac{x_C - x_B}{y_B - y_C}(x^* - x_s),\tag{4.4}$$

where

$$x_s = \frac{[x_C + r_{CB} \times x_B]}{1 + r_{CB}}, \quad y_s = \frac{[y_C + r_{CB} \times y_B]}{1 + r_{CB}},\tag{4.5}$$

and

$$r_{CB} = \frac{\delta_C^2 - \delta_B^2 + d(\mathbf{AP}_C, \mathbf{AP}_B)^2}{\delta_B^2 - \delta_C^2 + d(\mathbf{AP}_C, \mathbf{AP}_B)^2},\tag{4.6}$$

and the coordinates (x^*, y^*) will be given by the solution to Equations (4.3) and (4.4).

As a way to illustrate algorithm feasibility, we consider a linear scenario with only two APs. Let us say that $\mathbf{AP}_A = (x_A, y_A) = (0, 0)$ and $\mathbf{AP}_B = (x_B, y_B) = (x_B, 0)$, assuming that $\mathbf{n}_0$ lies between $\mathbf{AP}_A$ and $\mathbf{AP}_B$ and overestimates δ_A, and that δ_B of $d\langle\mathbf{AP}_A, \mathbf{n}_0\rangle$ and $d\langle\mathbf{AP}_B, \mathbf{n}_0\rangle$, respectively, define a triangle vertex. The geometric location estimate becomes

$$\frac{x_B}{2} + \frac{\delta_A - \delta_B}{2} \cdot \frac{\delta_A + \delta_B}{x_B},\tag{4.7}$$

while the maximum likelihood (ML) estimator for exponential additive error is

$$\frac{x_B}{2} + \frac{\delta_A - \delta_B}{2}.\tag{4.8}$$

Note that both estimators become closer as

$$\frac{\delta_A + \delta_B}{x_B} \to 1.$$

Simulations show that results for the ML and geometric estimator are fairly close (Figure 4.6), and as a matter of fact, in a scenario where distance measures are noisy, the geometric estimator is less than 3% less accurate than the ML estimator for exponential noise, with a mean error lower than 20% of the true value. That is,

$$E\left\{\frac{\delta_A}{d\langle\mathbf{AP}_A, \mathbf{n}_0\rangle}\right\} \leqslant 1.2.$$

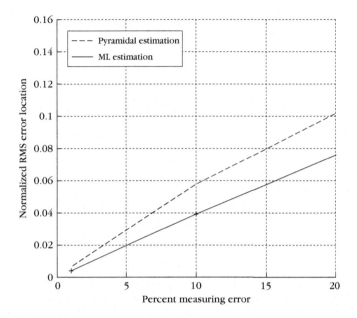

FIGURE 4.6

ML and pyramidal estimation performance versus percent measuring error.

Estimation Accuracy

To assess the accuracy of the pyramidal estimator, in the planar scenario, the Euclidean distance between the estimate and the actual location is adopted as a performance criterion. We study the estimation error ε and its dependency on the accuracy of distance measurements $d(P_{Ak-1}, P_{Ak})$ as well as on the way the paths are established.

Connecting paths exhibit spread that is influenced by the node density, reachability radius, and routing algorithm. For instance, at the i-th link, the hop is meant to occur in the direction Φ_i to the destination node. However, due to the random distribution of nodes, the hop may take place in the direction $\Phi_i + \varphi_i$ where $\{\varphi_i\}$ are considered to be independent and identically distributed random variables. Under the assumption of closest-to-destination node selection, φ_i behaves much like a zero-mean Gaussian distributed random variable with standard deviation (σ_φ) that increases as the node density grows.

The PL algorithm performance is assessed under a homogeneous scenario assumption, considering that subscribers are randomly distributed in an area defined by three fixed APs evenly placed in a circumference of radius R_{MAX} that define the networking area (see Figure 4.7).

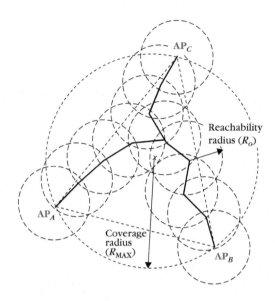

FIGURE 4.7

Pyramidal coverage area.

Although in a planar Poisson scenario, the k-th closest node to any reference point tends to follow a gamma distribution [11], in practice, the node reachability in the ad hoc environment is limited by the available transmitting power and sensitivity constraints that set a reachability radius R_0. Beyond this, no link can be set.

To locate a node, the paths linking it to the APs are set considering the reachability radius and the closest-to-destination node selection and random number of links in a path will be proportional to end-node separation. To assess the algorithm performance, simulations were conducted and reported results involve 10^4 random realizations of location estimates for each scenario.

The error analysis is reported in a normalized form as

$$\mu = \frac{\varepsilon}{R_{MAX}},$$

where ε is the distance between the estimated location and true location. Errors tend to be larger for small node densities.

Normalized error distribution $F_\mu(\xi)$ is presented in Figure 4.8a for different numbers of nodes in the coverage area (ρ). It can be seen that location estimation performance improves with node density; for instance, for 500 nodes in the networking area the location error is smaller than $0.1\ R_{MAX}$ in most cases. While for 100 nodes in the coverage area, only about 45% of the cases are below

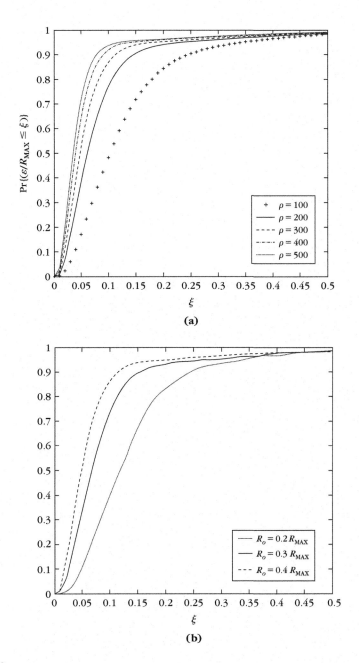

FIGURE 4.8

Pyramidal error performance. (a) Effect of number of nodes on normalized error distribution.
(b) Effect of reachability radius R_o on normalized error distribution.

$0.1 R_{MAX}$. In this simulation, the reachability radius was set as $R_o = 0.4 R_{MAX}$. Figure 4.8b shows the impact of the reachability radius relative to the coverage radius.

Note that a reachability radius of $0.4\, R_{MAX}$ leads to a location error smaller than $0.15\, R_{MAX}$, while for a reachability radius of $0.2\, R_{MAX}$ only about 60% of the results are below $0.15\, R_{MAX}$.

Up to this point, location errors are induced only by the path selection process, as no error in range estimation has been considered. However, it has been stated that a realistic scenario is subject to propagation impairments that cause range estimation error, so the distance overestimates are of the form:

$$\delta_A = \sum_{k=1}^{z} \left[d\left(P_{A(k-1)}, P_{Ak}\right) + \Delta_k \right], \tag{4.9}$$

where $\Delta_k, k = 1 \dots z$, denotes the range measurement error associated with the k-th hop. Travel measurements relate to first arrivals that often exhibit an exponential behavior [12]. Δ_k and Δ_p are considered to be independent and equally distributed for all k and p. For simulation purposes, the mean range error $E\{\Delta_p\}$ was taken as a factor γ of the actual distance $d\left(P_{A(p-1)}, P_{Ap}\right)$ (i.e., $E\{\Delta_p\} = \gamma \cdot d\left(P_{A(p-1)}, P_{Ap}\right)$.

Assuming a density of 300 nodes per covered area, Figure 4.9 shows how location error further degrades with distance the estimated error for 300 nodes.

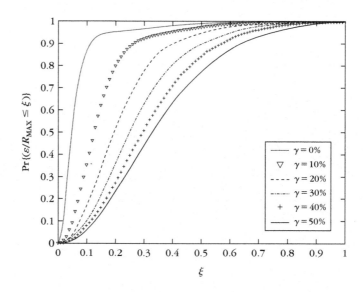

FIGURE 4.9

Normalized error distribution for different error factor γ; 300 nodes per coverage area.

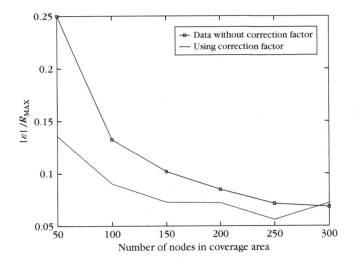

FIGURE 4.10

Impact of correction factor on normalized error versus number of nodes.

Recalling that $d\langle \mathbf{AP}_A, \mathbf{n}_0 \rangle \leq \delta_A$, we can approximate $d\langle \mathbf{AP}_A, \mathbf{n}_0 \rangle = \zeta \cdot \delta_A$, where ζ is a distance-correction factor. This coefficient will depend on the node density through the spread of the paths. Taking advantage of the fact that the true distances between APs are known, the ζ value can be estimated by comparing the addition of hop lengths interconnecting APs versus the actual AP separation. Figure 4.10 shows the improvement due to introduction of the correction factor ζ for a scenario of 100 nodes per coverage area.

4.2.3 Manhattanized Algorithms

Discretization is a common practice that provides certain advantages at the expense of introducing quantization errors. As in ad hoc networks, the number of hops used to establish a connection is an integer number; it may be intuitive to discretize the location space, considering the number of hops. In this section, several location techniques applicable to quantized scenarios are considered.

Let us assume that the location space is conceptually parceled by a Manhattan grid and that nodes will occupy any Manhattan crossing [13–15]. Thus, a location can be referred to in terms of the number of Manhattan steps that a node is from a location reference AP. And it can be considered that any location (x, y) is mapped into a k-dimensional vector $\boldsymbol{\delta} = (\delta_A, \delta_B, \ldots, \delta_K)$, where k is the number of APs and δ_j is the Manhattan separation of the node from the $\mathbf{AP}_j$ (Figure 4.11).

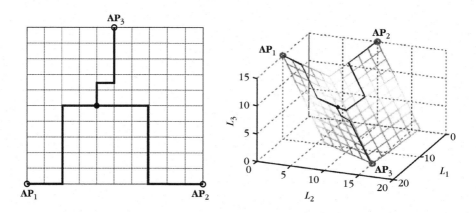

FIGURE 4.11

2D to 3D location mapping.

Note that the mapping $(x, y) \leftrightarrow (\delta_A, \ldots, \delta_K)$ is biunique; therefore (x, y) can be recovered from the estimated distances $\{\delta_i\}$. Since in practice, the location of the mobiles may depart from the corners of the Manhattan grid, a quantizing error is introduced and the suitability of this method depends, to a large extent, on the adequate estimation of the Manhattan distances.

As a first approach for Manhattan distance estimation in an ad hoc scenario, we assume that each hop corresponds to a unitary Manhattan step. This provides very rough Manhattan distance estimates. However, they can be improved by multiplying the number of hops by the appropriate weight coefficient S_f.

The scaling factor S_f can be obtained by comparing the number of hops in a path connecting two known location nodes (as APs for instance) against the actual Manhattan separation of the APs. This process can be repeated and averaged for all available paths interconnecting known location sites with known locations so that S_f designates the mean scaling factor.

That is, a node N_i that hops away from $\mathbf{AP}_i$ will be said to be at a Manhattan distance $\delta_i = \lceil S_s \times N_i \rceil$, where notation $\lceil \chi \rceil$ represents the smallest integer no lower than χ.

An alternative for hop-length estimation may rely on field strength and delay propagation measurements. When these measurements are available, they can be useful to calculate the Manhattan distance. Let us consider a node at location $\mathbf{P}_o$ that is linked to a node at location $\mathbf{P}_1$, which can be at most R units apart (R is the radio reachability range). In Figure 4.12a, it can be seen that depending on the region where $\mathbf{P}_1$ is, it can be assigned a Manhattan distance. For instance, points in $\mathcal{R}_2$ are closer to a Manhattan vertex that is two Manhattan units from P_0. Region $\mathcal{R}_1$ corresponds to points that are one unit apart from P_0, and similarly, $\mathcal{R}_0$ defines the region of points closer to P_0. Therefore, the hop or arch

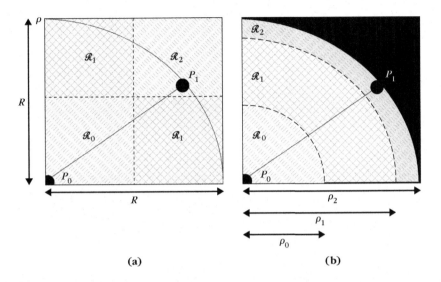

(a)   (b)

FIGURE 4.12

Range-dependent Manhattan distance estimation.

$\langle P_0, P_1 \rangle$ will be assigned a Manhattan distance according to the region where P_1 is located.

Although the precise P_1 location is unknown, the Euclidean distance $d(P_0, P_1)$ can be estimated through field strength and/or delay measurements, and the link $\langle P_0, P_1 \rangle$ can be assigned a Manhattan distance as follows. If $P_1 \in \mathcal{R}_i$, where $i = 0, 1, 2$, then Manhattan distance assignment will be $\delta(P_0, P_1) = i$ (see Figure 4.12b). Note that regions $\mathcal{R}_i$ are redefined as $R_i = \{P_1 : \rho_{i-1} \leqslant d(P_0, P_1) < \rho_i\}$ where thresholds $\{\rho_i\} i = 0, 1, 2$ are set in order to minimize Manhattan distance–assignment error according to a maximum likelihood criterion and $\rho_{-1} = 0$.

Other distance estimation improvements can be obtained, noting that the Manhattan layout is immaterial. Therefore, the basic Manhattan square can be smaller than that defined by the reachability radius R. For instance, the Manhattan basic step can be set as $\frac{R}{m}, m > 2$. This allows reduction of the quantizing error at the expense of increasing the computational effort.

In Figure 4.13, we illustrate how regions $\mathcal{R}_i$ are redefined for a Manhattan step given by $\frac{R}{3}$. As the Manhattan step becomes smaller, the number of assignment regions $\mathcal{R}_i$ increases and it is possible to obtain a more accurate representation of the route linking every AP with the mobile to locate.

Once the Manhattan distance estimates are available, several location algorithms can be invoked and the error performance will also depend on other parameters such as node density and routing strategy. One way of characterizing

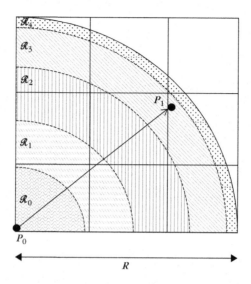

FIGURE 4.13

Finer Manhattan step ($L = R/3$).

these environments is through the directivity of the path link or angular mismatch of the hop direction and the intended direction.

At the i-th stage in the route, a hop is meant to occur in the direction ϕ_i to the destination node. However, due to the random distribution of nodes, the hop may take place in the direction $\phi_i + \varphi$. The random variable φ can be broadly characterized by its maximum values or its variance. For the purposes of this study, we assume that φ is uniformly distributed in the interval $|\varphi| \leqslant \varphi_{MAX}$. If φ_{MAX} is small, we have a *direct routing environment* (DRE), while if φ_{MAX} is large, we have a *spread routing environment* (SRE). In this study, we name those cases with $\varphi_{MAX} \leqslant 1.0472$ radians as DRE, while SRE corresponds to $\varphi_{MAX} \geqslant 2.0944$ radians; otherwise, we have a *typical routing environment* (TRE). φ_{MAX} will be referred to as the spreading index.

Manhattan Trilateration

In congruence with the proposed discretization, the Manhattan distance metric of two nodes $\mathbf{n}(x_n, y_n)$ and $\mathbf{m}(x_m, y_m)$ is defined as $d(\mathbf{n}, \mathbf{m}) = |x_n - x_m| + |y_n - y_m|$; thus, the intuitive conceptualization of the space changes [14]. For instance, a Manhattan circumference $MC_{AP_i}(\delta_i)$ of radius δ_i with a center at $\mathbf{AP}_i$ resembles a diamond (see Figure 4.14), according to

$$MC_{\mathbf{AP}_i}(\delta_i) = \{(x, y) : |x - x_i| + |y - y_i| = \delta_i\}. \tag{4.10}$$

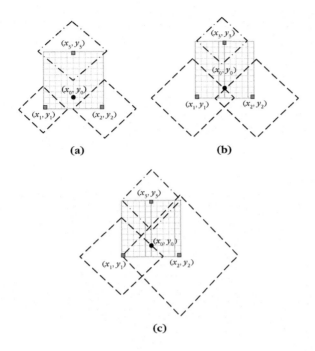

FIGURE 4.14

Nonintersecting Manhattan circles scenarios.

Assuming that a node to be located can establish ad hoc connections to the nearest three APs, the triplet $(\delta_1, \delta_2, \delta_3)$ where δ_i denotes the Manhattan distance between the mobile, and $\mathbf{AP}_i$ enables estimation of the position (x_0, y_0) as the intersection of the Manhattan circumferences

$$\bigcap_{i=1}^{3} MC_{\mathbf{AP}_i}(\delta_i).$$

Since δ_i is not the actual distance but an estimate, several ambiguities could occur. For instance, the "Manhattan circles" may not have a single intersection point as desired. In this case, the centroid search is invoked.

Other difficulties appear when

$$\bigcap_{i=1}^{3} MC_{\mathbf{AP}_i}(\delta_i) = \phi,$$

that is, when all circumferences do not intersect. In this case, the radii of the nonoverlapping circles are incremented according to $\delta_i = \delta_i + 1$ until the three circles overlap. Whenever three circles overlap, the point in the Manhattan grid closest to the centroid of the "circles" intersections is selected as the position location (x^*, y^*) of the node.

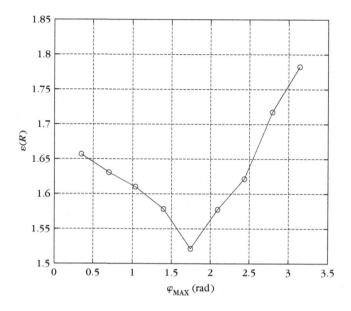

FIGURE 4.15

Mean square error versus maximum spreading index.

Results for these location strategies are shown in Figure 4.15, where the obtained Euclidean error is normalized to the Manhattan step. Results show that highly dispersive routes tend to produce large errors. This is because the path tends to overestimate Manhattan distances. Errors for low spreading indexes are associated with the quantizing process.

Vector Projection Algorithm

In the PL information (PLI) problem, a subscriber demands location determination while AP_i coordinates (x_i, y_i) are considered already known and available. It is also assumed that the routing algorithm can always establish connections from the subscriber to the nearest three APs. Thus, a mapping of the form $(x_o, y_o) \rightarrow (\delta_A, \delta_B, \delta_C)$ occurs, where δ_i denotes the Manhattan distance between the mobile and AP_i, $i = A, B, C$. Note that the mapping $(x_o, y_o) \rightarrow (\delta_A, \delta_B, \delta_C)$ is biunique; therefore, (x_o, y_o) can be recovered from the observed distances $(\delta_A, \delta_B, \delta_C)$. In practice, the location of the mobiles may depart from the corners of the Manhattan grid [14]. The closer the representative Manhattan path length is to the actual Manhattan separation, the better the location estimates may be.

Although the precise value of δ_i is not known, the number of hops Δ_i in the path linking the mobile and corresponding AP_i can initially be used as a length estimate. Quantization errors as well as inaccuracies due to erratic trajectories will occur.

Length inaccuracies may prevent the application of a direct inverse mapping process. However, this problem can be tackled by a vector projection algorithm (VP) as follows. Path inaccuracies may produce mismatching between the estimated Manhattan distance Δ_i and the actual distance value ($\Delta_i = \delta_i + \varepsilon_i$, where ε_i stands for a distance estimation error); PL estimation is not possible when the evidence $(\Delta_A, \Delta_B, \Delta_C)$ is ouside the $(x_o, y_o) \rightarrow (\delta_A, \delta_B, \delta_C)$ mapping.

However, estimated Manhattan distances $(\Delta_A, \Delta_B, \Delta_C)$ can be projected onto the closest point $(\delta_A, \delta_B, \delta_C)$ in the mapping surface, such that the inverse mapping $(\delta_A, \delta_B, \delta_C) \rightarrow (x_o^*, y_o^*)$ can be conducted.

As discussed previously, estimate accuracy depends on the prevailing routing algorithm scenario, which can be classified according to the path's dispersive characteristics. Simulation results are reported for various routing environments (spread index). Figure 4.16a presents results for the basic vector projection scheme and considers the impact of the correcting factor. Figure 4.16b considers the value of using range estimates, while Figure 4.16c deals with range estimates plus finer discretization levels.

Euclidean distance errors between the actual coordinates (x_0, y_0) and the inferred location (x_0^*, y_0^*) have been adopted as goodness criteria. The reported error is normalized to the basic Manhattan step. Simulations were conducted for a homogeneous scenario where subscribers are randomly distributed in a square area containing three fixed APs.

We draw attention to the fact that the Euclidean error ε may increase if the mean number Δ_i of hops in the routes linking the subscriber with the APs grows. The mean number of hops per path, while depending on spread routing characteristics, will also be a function of total operation area size. The basic number of nodes in the coverage area is determined by the transmission range R of the mobiles; for a square area A, we expect to have $(N + 1)^2$ nodes where

$$N = \sqrt{A}\big/R.$$

Figure 4.16d shows the Euclidean distance error ε for the vector projection algorithm as a function of the size of the network operation area referred to in terms of N, which relates to the mobile's transmission range R. Simulations were conducted for the three routing environments (DRE, TRE, SRE) defined in previous paragraphs. It can be said that the smaller the operation area, the smaller the distance error ε will be. In addition, better results are obtained for more directive routing paths.

The network operation area can vary depending on the geographic position of the nearest available APs enabled to execute the PL process. The nodes in the

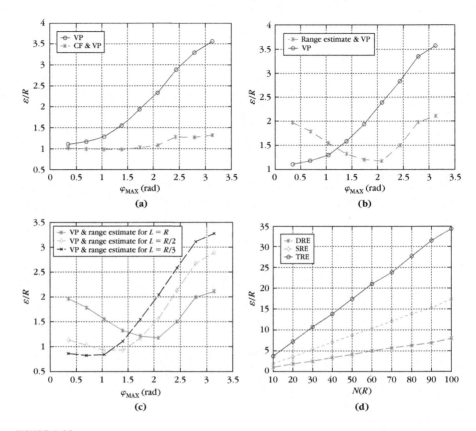

FIGURE 4.16

Vector projection (VP) algorithm. (a) Impact of the correcting factor (CF). (b) Impact of the range estimate for a Manhattan step defined by $L = R$. (c) Impact of the finer Manhattan step and range estimate. (d) Euclidean distance error versus size of network operation in terms of Manhattan crossings N.

ad hoc network may have diverse capabilities; for instance, some may have GPS features. This fact can be advantageous, as their locations could be used as fixed points, allowing size reduction of the considered Manhattan grid, reducing at the same time the error and computational effort.

A Linear Programming Approach

A commonly used metric in the separation of two locations is the distance $d(\mathbf{n}_i, \mathbf{n}_j) = [(x_i - x_j)^\alpha + (y_i, y_j)^\alpha]^{1/\alpha}$; $\alpha > 0$, where (x_i, y_i) denotes the coordinates of point $\mathbf{n}_j$. This generalized distance is known in the literature as the Minkowsky metric, which coincides with the Euclidean distance for $\alpha = 2$. For the discretized scenario, $\alpha = 1$ and it coincides with the already discussed Manhattan metric.

In the PL problem, a subscriber demands its location (x_0, y_0) to be known, while $\mathbf{AP}_i$ coordinates (x_i, y_i) have been considered to be available. Assuming that the ad hoc connections to the nearest three network APs have been established, a triplet $(\delta_A, \delta_B, \delta_C)$ can be defined, where δ_i denotes the Manhattan distance between the mobile and $\mathbf{AP}_i$. We recall that Manhattan distances were estimated from the number of hops in the links and several estimation improvements, as finer discretization, correcting factors, and so on, can also be in place.

We assume, for a while, that the distance triplet $(\delta_A, \delta_B, \delta_C)$ is noiseless. The δ_i denotes the actual Manhattan separation, and the APs' location can be considered to define a coverage region where the mobile can be located. The coverage region can be subdivided into two subregions R_i, $i = 1, 2$ (as illustrated in Figure 4.17). This allows estimating the PL of a mobile falling in the Manhattan grid by observing relationships imposed by the grid itself.

For instance, if $\mathbf{P}_o$ is located in the region R_1, then its coordinates must satisfy

$$x_0 = \frac{x_C - [(\delta_B + \delta_C) - d_{BC}]}{2}, \quad y_0 = \frac{[(\delta_A + \delta_B) - d_{AB}]}{2}. \qquad (4.11)$$

On the other hand, if the node is in region R_2, then the location of the node is at

$$x_0 = \frac{x_C - [(\delta_A + \delta_C) - d_{AC}]}{2}, \quad y_0 = \frac{[(\delta_A + \delta_B) - d_{AB}]}{2}. \qquad (4.12)$$

Note that d_{ij} denotes the true Manhattan separation of $\mathbf{AP}_i$ and $\mathbf{AP}_j$. To determine the applicable subregion R_i, $i = 1, 2 \ldots$. As seen in Figure 4.17, nodes

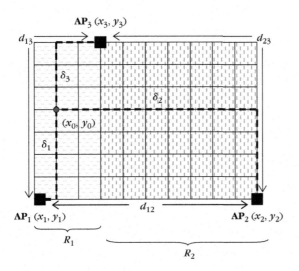

FIGURE 4.17

Operation region defined by AP location.

in R_1 are closer to $\mathbf{AP}_A$ than to $\mathbf{AP}_B$. Thus, the difference $\delta_A - \delta_B$ is compared to a threshold ξ that defines the size region, and ξ is conveniently defined from coordinates of $\mathbf{AP}_C$.

To work with a single region, coordinates of APs can notationally be modified as follows. Since the Manhattan layout is immaterial, $\mathbf{AP}_A$ (x_A, y_A) and $\mathbf{AP}_B$ (x_B, y_B) are said, without any loss of generality, to have the same ordinate (i.e., they define the abscise axis). Coordinates of $\mathbf{AP}_C$ (x_C, y_C) will be modified to $(x_C + \Delta_{xC}, y_C)$, where the shift $\Delta_{xC} = x_A - x_C$ is a known quantity, as all the AP coordinates are available. This allows working with a single region as illustrated in Figures 4.18a and b.

Naming d_x the Manhattan separation among APs $\mathbf{AP}_A$, and $\mathbf{AP}_B$, and d_y the Manhattan separation between APs $\mathbf{AP}_A$ and $\mathbf{AP}_C$, the location (x_0, y_0) can (in the absence of noise) be inferred from the distance in $\mathbf{AP}_A$, $\mathbf{AP}_B$, and $\mathbf{AP}_C$ to any node in the operation area $(\delta_A, \delta_B, \delta_C)$:

$$x_0 = \left[(\delta_B + \delta_C) - d_y\right]/2, \quad y_0 = \left[(\delta_A + \delta_B) - d_x\right]/2. \tag{4.13}$$

In a practical scenario, nodes do not necessarily lie on the corners of the Manhattan grid. In addition, due to the randomness of the node location and routing algorithms, connection paths tend to follow erratic trajectories. Therefore, direct application of Equation (4.13) renders in error. An integer search algorithm can be implemented to obtain the best performance. For the case of three APs, it must be noted that triplets $(\delta_A, \delta_B, \delta_C)$ associated with the Manhattan grid crossings must satisfy not only Equation (4.13) but a set of linear constraints that define a feasibility region. The feasibility region $\mathcal{F}$ is defined by the triplets $(\delta_A, \delta_B, \delta_C)$ that satisfy the following relations:

$$0 \leq |\delta_A - \delta_B| \leq d_x; \quad 0 \leq |\delta_A - \delta_C| \leq d_y; \quad \delta_A + \delta_C = d_x + d_y. \tag{4.14}$$

Therefore, the PLI acquisition can be considered equivalent to finding, in the feasible region, the triplet $(\delta_A^*, \delta_B^*, \delta_C^*)$ closest to the observation $(\Delta_A, \Delta_B, \Delta_C)$. Thus, the PLI acquisition becomes a linear integer-programming problem of the form

$$\min_{(\delta_A^* \delta_B^* \delta_C^*) \in \mathcal{F}} \left\{ |\delta_A^* - \Delta_A| + |\delta_B^* - \Delta_B| + |\delta_C^* - \Delta_C| \right\}. \tag{4.15}$$

Once $(\delta_A^*, \delta_B^*, \delta_C^*)$ is found, the location (x_0, y_0) can be obtained by application of Equation (4.13).

It is recognized that an increase in the number of APs provides additional evidence that may result in better performance. Thus, in a scenario with four APs, the previous procedure can be extended by resetting the AP coordinates to leave a single working region. For instance, locations $\mathbf{AP}_A$ and $\mathbf{AP}_B$ are kept as fixed references in the Manhattan grid, and the $\mathbf{AP}_C$ coordinates are shifted to have the same abscise as $\mathbf{AP}_B$ and the same ordinate as $\mathbf{AP}_D$. Similarly, $\mathbf{AP}_D$ will share the same abscise as $\mathbf{AP}_A$. This is illustrated in Figures 4.18c and d where the

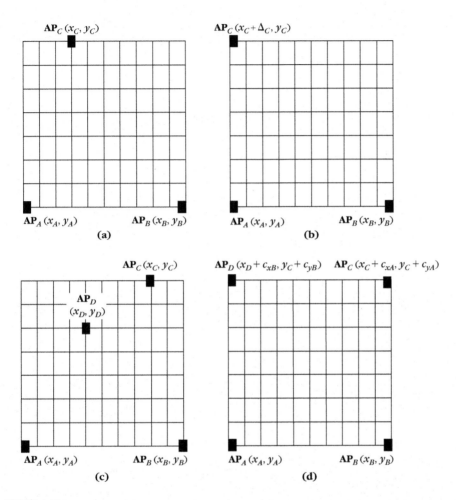

FIGURE 4.18

AP reallocation scenarios. (a) Original three-AP setting. (b) Reallocated three-AP setting. (c) Original four-AP setting. (d) Reallocated four-AP setting.

APs are meant to be reallocated at (x_A, y_A), (x_B, y_B), $(x_C + \Delta_{xC}, y_C + \Delta_{yC})$, and $(x_D + \Delta_{xD}, y_D + \Delta_{yD})$ where $\Delta_{xC} = x_B - x_C$, $\Delta_{yC} = y_D - y_C$, and $\Delta_{xD} = x_A - x_D$ are known constants.

When the four distances $(\delta_A, \delta_B, \delta_C, \delta_D)$ are obtained, the feasibility region $\mathcal{F}$ is defined by the distances $(\delta_A, \delta_B, \delta_C, \delta_D)$ that satisfy the following relations:

$$0 \leq |\delta_A - \delta_B| \leq d_x; \;\; 0 \leq |\delta_A - \delta_C| \leq d_y; \;\; \delta_A + \delta_C = d_x + d_y,$$
$$0 \leq |\delta_A - \delta_D| \leq d_y; \;\; 0 \leq |\delta_C - \delta_D| \leq d_x; \;\; \delta_B + \delta_D = d_x + d_y, \tag{4.16}$$

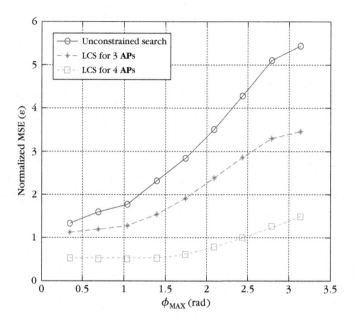

FIGURE 4.19

Performance of the linear constrained searching algorithm.

and the PLI acquisition becomes a linear integer-programming problem of the form

$$\min_{(\delta_A^* \delta_B^* \delta_C^* \delta_D^*) \in \mathcal{F}} \left\{ \left| \delta_A^* - \Delta_A \right| + \left| \delta_B^* - \Delta_B \right| + \left| \delta_C^* - \Delta_C \right| + \left| \delta_D^* - \Delta_D \right| \right\}. \qquad (4.17)$$

Once $(\delta_A^*, \delta_B^*, \delta_C^*)$ is found, the location (x_o, y_o) can be obtained by application.

Although the minimization process can be conducted without recourse to the feasibility region constraints, these allow a substantial improvement. This is shown in Figure 4.19 where unconstrained search is conducted for different routing spread scenarios, and results are compared with those of a restricted search for the three-AP case. Improvement due to the use of an additional AP is also shown in the figure.

Performance will depend on the accuracy of the Manhattan distance estimates. These could be improved by incorporating range measurements. Additional improvements include the use of finer discretization as discussed in previous sections. Thus, the Manhattan step can be reduced from the reachability radius R to R/m (m integer). The advantage of this discretization is shown in Figures 4.20a and b, for both the three- and four-AP scenarios.

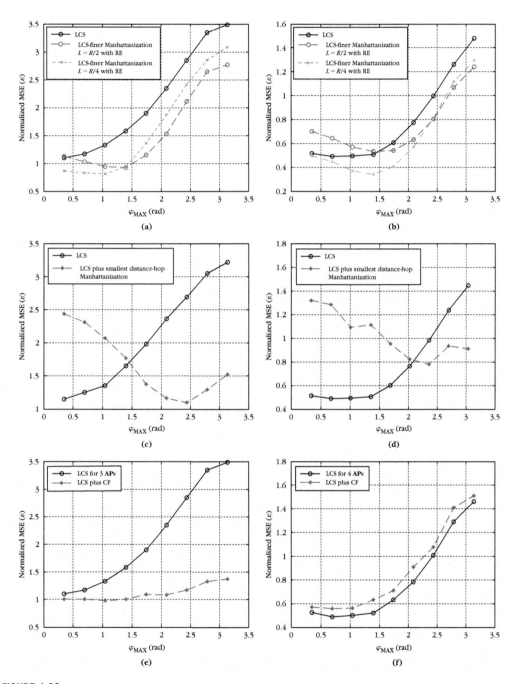

FIGURE 4.20

Error for LCS with range estimates. Finer Manhattan steps ($L = R/m$, $m = 2, 4$). (a) Three APs. (b) Four APs. Manhattanization based on smallest-distance hop. (c) Three APs. (d) Four APs. Correcting factor impact on LCS analysis. (e) Three APs. (f) Four APs. (LCS, linear constraint search).

Furthermore, based on the premise that the Manhattan grid is immaterial, a dynamic discretization can be adequate. In this case, the basic Manhattan step can be equated to the smallest distance hop η in the routes linking the three APs to the mobile. All the distances $d_{j,j+1}$ corresponding to the link connecting the nodes p_j and p_{j+1} are examined, and $\eta = \min_j \{d_{j,j+1}\}$ will be set as the basic Manhattan step so that hop lengths $d_{k,k+1}$ will be assigned Manhattan distances given by

$$\Delta_k = \frac{d_{k,k+1}}{\eta},$$

and the Manhattan path length δ_i will become

$$\delta_i = \sum_k \Delta_k$$

for all links in the path linking the node (x_0, y_0) to $\mathbf{AP}_i$. Note that the Manhattanization grid will be different for every case. This is the opposite of the previous case, where the same Manhattanization was conducted for all instances. Results for this strategy are presented in Figures 4.20c and d.

The average correction factor was shown to provide for the compensation of distance estimation errors. The impact of the correcting factors in the current location strategy is also shown for different spreading indexes both for three- and four-AP scenarios (see Figures 4.20e and f).

Three-Dimensional Manhattanized Case

Although most location problems considered in the literature pertain to planar scenarios, current developments in sensor networks as well as "vertical cities" scenarios demand 3D location strategies. The discretized scenarios can be readily extended to 3D scenarios. In this case, the reachability radius defines a sphere that will serve as the basis for Manhattanized 3D space where the node $\mathbf{n}_0$ to be located is contained. It will also be assumed that paths linking $\mathbf{n}_0$ to the APs can be established. As an illustration, let us consider a cubic space where APs are placed on opposite corners (see Figure 4.21). Let $\mathbf{n}_0$ be a node to be located that is δ_A Manhattan units from $\mathbf{AP}_A$ and δ_B units from $\mathbf{AP}_B$; APs $(\mathbf{AP}_A, \mathbf{AP}_B)$ are δ_{AB} units apart. If $\delta_A + \delta_B = \delta_{AB}$, $\mathbf{n}_0$ will be known to be in the cube face containing $\mathbf{AP}_A$ and $\mathbf{AP}_B$. In contrast, if $\delta_A + \delta_B \geqslant \delta_{AB}$ and in the absence of noise, $\mathbf{n}_0(x_0, y_0, z_0)$ will be known in a parallel plane with a y_0 ordinate:

$$y_0 = \frac{\delta_A + \delta_B - \delta_{AB}}{2}.$$

A similar argument can be followed for cube faces containing APs $(\mathbf{AP}_A, \mathbf{AP}_C)$ and $(\mathbf{AP}_A, \mathbf{AP}_D)$, so that the remaining coordinates will be

$$x_0 = \frac{\delta_A + \delta_D - \delta_{AD}}{2} \quad \text{and} \quad z_0 = \frac{\delta_A + \delta_C - \delta_{AC}}{2}.$$

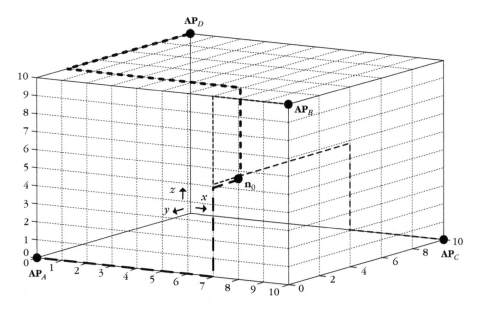

FIGURE 4.21

3D Manhattan location scenario.

Note that in the described process example, location is obtained using a single plane per coordinate. However, there are six planes that can be used for coordinate determination. This multi-evidence scheme will be useful in noisy environments in such a way that coordinates calculation leads to multiple solutions and the final proposed location will be the average location.

Figure 4.22 shows the error normalized against the basic Manhattan step. The simulation assumes a uniform distributed distance error. The figure shows results for a nonredundant scheme (one plane per coordinate) and a full-evidence scenario.

4.2.4 Relational and Fuzzy Approach

From a philosophical perspective, location makes sense only in relation to other locations. In other words, it is always necessary to establish a reference frame that allows description of location sites. This referential frame in some cases consists of coordinate systems that refer all location sites to an agreed-on origin. Examples are the rectangular and polar coordinate systems (among others) with a selection that is determined in practical terms as well as the origin of coordinates.

In some applications, origin is set on a convenience basis. For instance, in navigation a particular reference known as a fix is adopted and locations refer to that fix. When that location is too distant or uncertainty has increased, a new

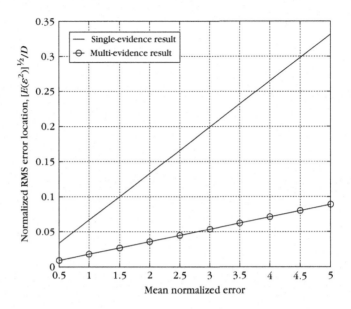

FIGURE 4.22

Normalized root mean square error.

fix is established. It may be convenient while adopting a new fix to retain the previous references. Another common practice is having a set of fixed locations shared by users so that the current fix is always selected from the reference set.

Trilateration has been discussed in previous chapters, and in this chapter we assume that node $n_0 = (x, y)$ in an ad hoc network is reachable from at least three APs $(AP_i(x_i, y_i), i = A, B, C)$ so that a trilateration process can be invoked.

In the ad hoc network, the links may have different lengths and we denote the mean hop length as ζ. Assuming that the node $n_0 = (x, y)$ to be located is N_i hops away from $AP_i(x_i, y_i)$, it can be considered $\delta_i = N_i \cdot \zeta$. In practice, the distances δ_i may exceed the actual node to AP_i separation, and a more convenient approach is to adopt

$$\mathbf{cent}\left[\bigcap_{\forall i} C_{APi}(\delta_i)\right]$$

as the location of n_0, where $C_{AP}(\delta_i) = \left\{(x, y) \,\middle|\, (x - x_i)^2 + (y - y_i)^2 \leqslant \delta_i^2\right\}$ and the **cent**[A] operator denotes the centroid of the area A.

Note that in some cases, $C_{APi}(\delta_i)$ and $C_{APj}(\delta_j)$ may not overlap, and in this case intersection $C_{AP_i}(\delta_i) \cap C_{APj}(\delta_j)$ can be replaced by intersection

$$C_{AP_i}(D_{ij} - \delta_i) \cap C_{APj}(D_{ij} - \delta_j), \tag{4.18}$$

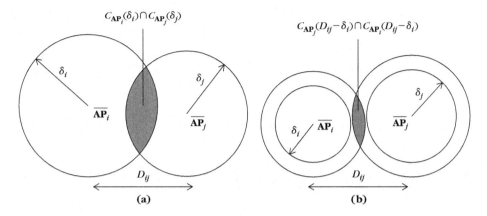

FIGURE 4.23

Coverage and separation of APs.

where D_{ij} is the separation on the $\mathbf{AP}_i$ and $\mathbf{AP}_j$ that is assumed to be known (see Figure 4.23).

Fuzzy Proportional Method

As discussed previously, in an ad hoc network the information about the nodes that a packet has traveled through is available. Also, a connected network is assumed. Thus, several algorithms based on the number of hops can be developed [9].

In a given scenario, node $\mathbf{n}_0$ to be located could be described by its estimated distance δ_i to the $\mathbf{AP}_i$. N_i and N_j are the minimum number of hops required to link node $\mathbf{n}_0$ to $\mathbf{AP}_i$ and to $\mathbf{AP}_j$, respectively. Node $\mathbf{n}_0$ can be assumed to be at a location on the segment connecting the APs maintaining the proportion

$$\frac{\delta_i}{\|\mathbf{AP}_i - \mathbf{AP}_j\|} = \frac{N_i}{N_i + N_j}. \tag{4.19}$$

Figure 4.24a illustrates connecting paths linking $\mathbf{n}_0$ to four APs. Figure 4.24b illustrates application of the Equation (4.19) on the lines connecting $\mathbf{AP}_A$ to $\mathbf{AP}_C$ and $\mathbf{AP}_B$ to $\mathbf{AP}_D$. Thus, for each pair of APs, the estimate coordinates (x, y) of node $\mathbf{n}_0$ will be given by

$$x = x_i + \frac{N_i}{N_i + N_j}(x_j - x_i) \text{ and } y = y_i + \frac{N_i}{N_i + N_j}(y_j - y_i). \tag{4.20}$$

Note that this scheme implies a normalization process with respect to APs separation. This process is repeated for all AP pairs, and the final location will be given by the centroid of the proposed solutions.

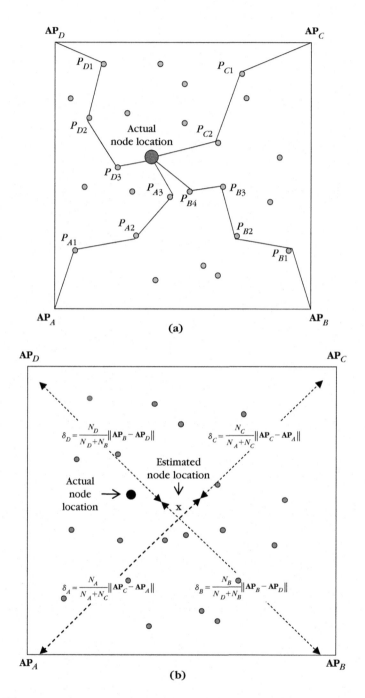

FIGURE 4.24

Fuzzy proportional scheme. (a) Paths linking node to APs. (b) Proportional node allocation.

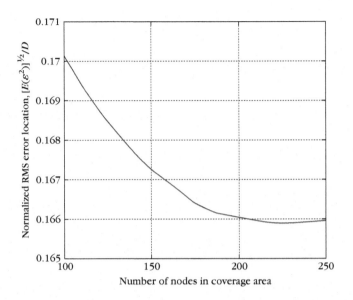

FIGURE 4.25

Fuzzy proportional performance as a function of nodes in coverage region.

Figure 4.25 gives error results for a scenario where the four APs are D distance units apart and they define a square area. Simulations were conducted for nodes generated with a spatial uniform process and a reachability radius R of $\frac{D}{\sqrt{50}}$. Results show that error tends to diminish as the number of nodes in the network increases.

Fuzzy Hyperbolic Algorithm

Note that relation

$$\frac{\delta_i}{\|AP_i - AP_j\|} = \frac{N_i}{N_i + N_j}$$

can be also translated into

$$\delta_j - \delta_i = \|AP_i - AP_j\| \cdot \left[1 - 2\frac{N_i}{N_i + N_j}\right], \tag{4.21}$$

which represents a hyperbola. Considering different APs pairs, the node $\mathbf{n}_0$ can be placed at the intersection of hyperbolas of the form

$$\sqrt{(x - x_j)^2 + (y - y_j)^2} - \sqrt{(x - x_i)^2 + (y - y_i)^2} = C_{ij},$$

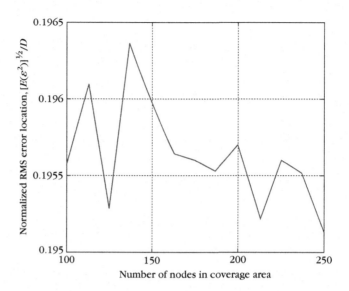

FIGURE 4.26

Fuzzy hyperbolic performance as a function of number of nodes in coverage region.

where constant C_y is given by

$$C_{ij} = \delta_j - \delta_i = \left\| \mathbf{AP}_i - \mathbf{AP}_j \right\| \cdot \left[1 - 2\frac{N_i}{N_i + N_j} \right].$$

Note that accuracy will depend on node density, the routing selection process, and δ_i estimation.

Similar to previous schemes, mismatches in difference estimations lead to different solutions. These ambiguities are sorted out by taking the centroid of those solutions contained in the feasibility region defined by the convex hull containing the APs of interest. Because the hyperbola intersection process produces multiple solutions, its performance is inferior compared to the proportional method. Results of both schemes based on the scenario discussed earlier are shown in Figure 4.26.

Minimization Using Rough Evidence

Another algebraic heuristic can be established upon considering that the node to be located $\mathbf{n}_0(x_0, y_0)$ will have associated distances $d_i = \sqrt{(x - x_i)^2 + (y - y_i)^2}$ to the $\mathbf{AP}_i(x_i, y_i)$.

Although d_i is unknown, its estimate δ_i can be inferred with the use of any of the proposed schemes. Thus, the node $\mathbf{n}_0$ location will be assumed to be roughly in a region where distances to $\mathbf{AP}_i$ are somehow close to the estimated

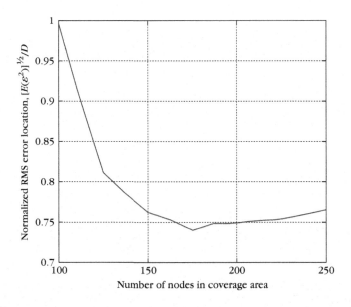

FIGURE 4.27

Simulation results for minimization based on rough evidence.

distances (see Figure 4.27). Therefore, the search will try to minimize differences $|d_i - \delta_i|$. In brief, $\mathbf{n}_0$ will be placed at an (x, y) point that minimizes the objective function

$$J(x, y) = \sum_i \left(\sqrt{(x - x_i)^2 + (y - y_i)^2} - \delta_i \right)^2. \tag{4.22}$$

The resulting error for different node densities is shown in Figure 4.27.

Neighborhood Method

Location is always described in relation to a coordinate referential system. These references are often considered fixed and known. In practice, we often have only an imprecise idea of the referential system, and we gather a location in the context of the relations to several vague descriptors. In this section, we deal with some location relational schemes without direct recourse to APs.

To estimate its coordinates, node $\mathbf{n}_0$ inquires of nodes in its neighborhood about their location. The R neighborhood of $\mathbf{n}_0$, $(N_R(\mathbf{n}_0))$, is defined by all the nodes reachable from $\mathbf{n}_0$ in a single hop. In brief, node $n_j \in N_R(\mathbf{n}_0)$ if $|n_j - \mathbf{n}_0| < R$ when R is the reachability radius.

Each node in an ad hoc network will be connected at least to a known location AP that provides connectivity to the rest of the network. A stamp can be added to a packet when it passes through a node. This labeling process allows identifying the nodes that the packet has traveled along, and the information is adequate to obtain a degree-vicinity matrix for each node, as described earlier. The vicinity degree of two nodes is the minimum number of hops required to establish a connection. The first vicinity matrix is defined by the neighborhoods $(N_R(n_j))$ and it will be coincident with adjacency matrix $\mathbf{A}_{n \times n} = \{a_{ij}\}$ illustrated in Figure 4.4a.

The second vicinity degree of a node n_i with the rest of the nodes can be expressed by a matrix ${}_2\mathbf{A}_{n \times n} = \{{}_2A_{ij}\}$ where self-vicinity is defined to be zero. This is ${}_2A_{ii} = 0$ and for $i \neq j, {}_2A_{ij} = 1$ if $\sum a_{ik} \times a_{kj} \neq 0$; ${}_2A_{ij} = 0$ otherwise. Similarly the third vicinity degree matrix ${}_3\mathbf{A}_{n \times n} = \{{}_3A_{ij}\}$ can be obtained from the second vicinity degree matrix product ${}_2\mathbf{A}_{n \times n} \times {}_2\mathbf{A}_{n \times n}$ where ${}_3A_{ii} = 0$ and for $i \neq j, {}_3A_{ij} = 1$ if $\sum_2 A_{ik} \times {}_2A_{kj} \neq 0$; ${}_3A_{ij} = 0$ otherwise. This procedure is extended to obtain other vicinity degrees of all nodes n_0.

In the proposed scheme, a node n_0 is presumed to obtain its location from the vicinity of its neighbors to reference APs. To describe a basic algorithm, we initially consider a linear scenario (see Figure 4.28a).

Let us assume that node n_0 queries its neighbors on their vicinity to $\mathbf{AP}_A$ and $\mathbf{AP}_B$, and these queries are sorted so that $N = N_A + N_B$ denotes the total number of n_0 neighbors, where N_A and N_B denote the number of neighbors closer to $\mathbf{AP}_A$ and $\mathbf{AP}_B$, respectively.

If $N_A = N_B$, it can be assumed that n_0 is located at the middle point on the line connecting $\mathbf{AP}_A$ and $\mathbf{AP}_B$. Under the assumption that neighbors spread evenly in the neighborhood, the proportion

$$\zeta = \frac{N_A}{N_A + N_B}$$

of the nodes closer to $\mathbf{AP}_A$ provides an indicator of the amount Δ_{AB} that n_0 is shifted toward $\mathbf{AP}_A$ [9]. Note that Δ_{AB} defines a cord location (see Figure 4.28a) in the reachability circle and equates

$$\zeta = \frac{B}{A + B},$$

ζ, and Δ_{AB} related through

$$B = \frac{R^2}{2} \left[\cos^{-1} \left(\frac{2\Delta_{AB}^2}{R^2} - 1 \right) - \sin^{-1} \left(\cos^{-1} \left(\frac{2\Delta_{AB}^2}{R^2} - 1 \right) \right) \right], \tag{4.23}$$

where A and B denote the area partitioned by the cord.

Note that in a planar scenario and under the same homogeneous hypothesis, the same proportion ζ is kept for all locations on the line $n_0 \perp_{AB}$ passing through n_0 and orthogonal to the line connecting $\mathbf{AP}_A$ and $\mathbf{AP}_B$ (see Figure 4.28b).

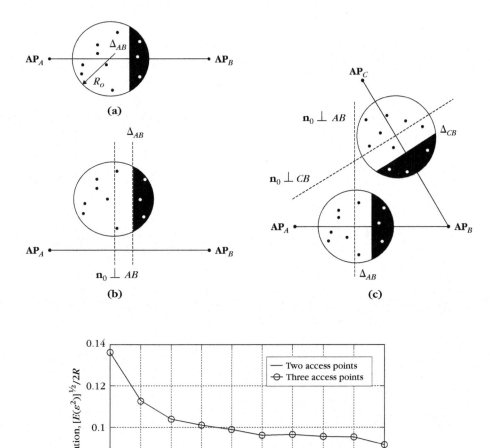

FIGURE 4.28

Neighborhood-based scheme. (a) Location drift dependent on number of neighbors.
(b) Drift invariance. (c) 2D location scheme. (d) Error performance as a function of
number of nodes.

The same process is applied to the pair ($\mathbf{AP}_B$ and $\mathbf{AP}_C$). The node can be considered located at the intersection of lines $\mathbf{n}_0 \perp_{BC}$ and $\mathbf{n}_0 \perp_{AB}$ (Figure 4.28c). The process can be extended for all possible AP pairs. However, intersection of the lines can occur at different points and the centroid of the intersections will be taken as the location of the node of interest.

Figure 4.28d shows results for the proposed algorithm for two and three APs. In many discussed scenarios, the error tends to diminish as the number of APs increases. This comparison is not applicable here, as the two-AP case corresponds to a linear scenario and the three-AP scheme is representative of a planar scenario.

Relative Distance Location

Position location information (PLI) is often obtained through trilateration processes involving known location reference points [16, 17]. However, in reconfigurable networks, such as ad hoc and sensor networks, these schemes cannot be directly applied since links might break and direct connectivity between fixed references and mobiles is not ensured [18]. To overcome this problem, nodes can establish relational tables according to the topology at the same time that procedures such as neighbor discovery and routing take place. These tables will position each node in relative locations according to reachable neighboring nodes by forming relational triangles where each node is a vertex of such triangles. These logical relationships can be used to estimate the relative position of the nodes. Although triangle construction schemes have been proposed (see [17]), construction cannot be guaranteed due to the mobility of the nodes and link availability.

In this section, a relational approach suitable for PLI acquisition in reconfigurable networks is presented. In relational networks, the entities are not defined by themselves, but by the type and degree of their relationship to other entities of the same class. Thus, when all the relations are defined, a global picture of the network is achieved. For instance, for a family of individuals, knowledge of the isolated family links may assist in building the family tree. And the more links that are known, the better the tree picture [19].

In a reconfigurable network, subscribers relay information to nearby neighboring customers, so that messages propagate from node to node until a gateway or access point connected to the rest of the network is reached, achieving in this manner the desired connectivity and functionality. The sets of neighboring link connections can help in the inference of node location on the network.

The routes are dynamically obtained, bearing in mind the prevailing network conditions. However, the routing algorithms in general are constructed underlying a minimal distance criterion or node separation. It is also assumed that at an AP, an arriving packet contains the identification of the sequence of nodes traversed. With this information, we establish a basic relational backbone that, together with distance estimates, provides for PLI estimation.

In Rao et al. [20], the positions of the nodes are obtained by an iterative method where some fixed nodes on the perimeter are used to establish the relationships. In our method, which is also iterative, the fixed nodes or APs do not need to be on the perimeter. Without loss of generality, one can consider APs at the perimeter or in the middle of the network.

In addition, the number of perimeter nodes, as noted in Rao et al. [20], is significantly higher than the number of APs we use. We also provide an algorithm with improved location-estimation precision. While the method in Rao et al. is used for routing support with the use of approximate positions, in the method presented here, the objective is to obtain a precise position of the nodes and compare those positions to the true locations. We implement a geometric process based on circle intersections to increase accuracy. The Rao et al. results show certain tolerance to location error from the point of view of routing, but it is also shown that for a lower number of nodes, whether for high- or low-density networks, their obtained locations are quite different from the true positions. We also would like to mention that in this section, we do not discuss the means by which information is exchanged, nor the way that topology is organized. All the nodes interact through the algorithm and participate in determining the final positions of the nodes.

Location Algorithm Description

The basic scenario considered is a square area where nodes are generated with a spatial Poisson process such as that shown in Figure 4.29, where nodes and existing links are illustrated. While $\mathbf{AP}_A, A = 1, \ldots, 4$, in the same figure are at known positions (x_A, y_A), respectively, remaining node locations are unknown and they will be estimated through the proposed algorithm.

It is assumed that the network is a connected graph. Thus, for any pair of APs, a path can be established. In our particular case, these paths are obtained by using a shortest-path algorithm such as Dijkstra's [21], but other routing algorithms and protocols could be used; thus, the presented method works independently of the routing used.

First Estimation

Recall that nodes make routing decisions based on topology information (i.e., neighboring nodes, reachable nodes, nodes within coverage areas, etc.), but not on precise location information. First, using the routing algorithm, we obtain paths between any pair of APs in the network. Assume that we look into

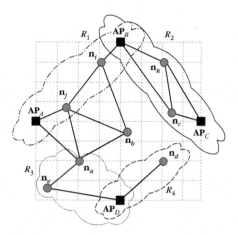

FIGURE 4.29

Example of reconfigurable network.

the path joining $\mathbf{AP}_A$ and $\mathbf{AP}_B$. Define $P_{AB}(k)$ as the k-th node in the path—that is, $\mathbf{AP}_A \leftrightarrow P_{AB}(1) \leftrightarrow \cdots \leftrightarrow P_{AB}(k) \leftrightarrow \cdots \leftrightarrow \mathbf{AP}_B$. Because we lack additional information at this stage, the nodes in the mentioned path can sequentially be assigned locations evenly placed on the segment $\langle \mathbf{AP}_A, \mathbf{AP}_B \rangle$. However, internode separation estimates for the existing links can be obtained by field-strength measurements or from propagation delay measurements [22, 23]. The addition of individual hop lengths is assumed to be proportional to the route length. This proportion is applied to scale at the individual hop placement. Thus, for instance, nodes $P_{AB}(k)$ and $P_{AB}(m)$, with coordinates $(x_{AB}(k), y_{AB}(k))$ and $(x_{AB}(m), y_{AB}(m))$, respectively, can be placed on the straight segment $\langle \mathbf{AP}_A, \mathbf{AP}_B \rangle$ as nodes $P_{AB}(k)^*$ and $P_{AB}(m)^*$ with coordinates $(x_{AB}^*(k), y_{AB}^*(k))$ and $(x_{AB}^*(m), y_{AB}^*(m))$, respectively. The separation distance between $P_{AB}(k)^*$ and $P_{AB}(m)^*$ is set keeping the proportion

$$d\langle P_{AB}(k)^*, P_{AB}(m)^* \rangle = \frac{L_{A,B} \cdot \delta_{k,m}}{\Lambda_{A,B}}, \qquad (4.24)$$

where $\delta_{k,m}$ denotes the measured segment or hop length $\langle P_{AB}(k), P_{AB}(m) \rangle$, $L_{A,B}$ stands for the actual $\langle \mathbf{AP}_A, \mathbf{AP}_B \rangle$ separation distance (straight line), and $\Lambda_{A,B}$ is given by the addition of total hop lengths resulting from measurements in the path that joins $\mathbf{AP}_A$ and $\mathbf{AP}_B$ [19]. Note that because $L_{A,B}$ is given by the length of the straight line joining $\mathbf{AP}_A$ and $\mathbf{AP}_B$, and the intermediate nodes are not necessarily on the straight line of the path, $\Lambda_{A,B} \geqslant L_{A,B}$, therefore $L_{A,B}/\Lambda_{A,B} \leqslant 1$. Thus, this proportion depends on the spread of the paths. The process can be extended to all routes connecting two APs. By getting the different paths joining APs, we can see that some nodes could be part of several of those paths.

As node $\mathbf{n}_k$ can belong to M different paths, some ambiguities may occur. In order to prevent multiple location assignments for the same node, we provisionally assume that its location is (x_k^*, y_k^*) by averaging all the positions obtained by each of the M paths to which the node belongs. That is,

$$x_k^* = \frac{1}{M} \sum_{\forall AB} x_{AB}(k) \quad \text{and} \quad y_k^* = \frac{1}{M} \sum_{\forall AB} y_{AB}(k). \tag{4.25}$$

In this way, a relational backbone network is established, providing the first PL estimate for all nodes on the network.

Relational Location Adjustments

As seen in the subgraph R_2 in Figure 4.29, some adjacent nodes with their connecting links define triangles, and they can be considered rigid bodies. Furthermore, triangle concatenations can be treated as a single rigid body, thus reducing uncertainty and computational complexity. Let us take, for instance, the triangle defined by vertices $\mathbf{AP}_C$, $\mathbf{n}_k$, and $\mathbf{n}_c$, where $\mathbf{AP}_C$ is a hard reference (as their coordinates are known) and a node (let us say $\mathbf{n}_k$) connected to $\mathbf{AP}_C$ can be used to define a provisional directional reference. Recall that the distance estimate between nodes $\mathbf{n}_k$ and $\mathbf{n}_c$ is defined as $\delta_{k,c}$. Because the distance estimate $\delta_{C,k}$ is available, the location (x_k^*, y_k^*) given by Equation (4.2) can be shifted along the line defined by nodes $\mathbf{AP}_C$ and $\mathbf{n}_k$, maintaining the measured separation between them—that is, $\delta_{C,k}$. Once (x_k^*, y_k^*) has been redefined, (x_c^*, y_c^*) can be adjusted to the nearest point whose distances to $\mathbf{AP}_C$ and $\mathbf{n}_k$ match the measurements $\delta_{C,c}$ and $\delta_{c,k}$, respectively. Thus, the triangle $\mathbf{AP}_C$, $\mathbf{n}_k$, $\mathbf{n}_c$ is adjusted, and we continue adding vertices and distance estimates adjacent to already defined triangles until the whole triangle concatenation in the network is completed.

When another AP becomes part of the concatenation, its estimated coordinates can depart from the actual location. This is because an uncertain location (x_k^*, y_k^*) was used to set a provisional and arbitrary directional reference. In addition, available distance estimates may depart from the actual separation due to propagation impairments. At this point, the triangle concatenation must be scaled and rotated with the center at $\mathbf{AP}_C$ so that coordinates of the other AP match those of the true site. These scaling and rotating procedures are conducted for all triangle concatenations containing two APs. In the absence of measurement noise, our algorithm leads to very close results to actual node locations for those nodes in the triangle concatenation. However, in some cases during the concatenation construction, triangle mirror dilemmas may appear. These can be resolved from the available backbone information or by observing the separation to other nearby nodes. Also, some triangle concatenation may not include two APs; for example, see the concatenation of the triangles in Figure 4.29 involving nodes $\mathbf{AP}_A, \mathbf{n}_j, \mathbf{n}_a, \mathbf{n}_b,$ and $\mathbf{n}_i$.

Noting that $\delta_{B,i}$ is directly obtained as a measurement, and $\delta_{A,i}$ can be estimated from the triangle concatenation, the position of node $\mathbf{n}_i$ can be obtained by carrying out an intersection of two circles, one centered on $\mathbf{AP}_A$ and another centered on $\mathbf{AP}_B$. It is pointed out that, while the algebraic process leads to two solutions (i.e., two overlapping circles with a nonempty intersection will have two points where their circumferences meet), the relational information allows selection of the more viable solution. A similar procedure can be applied to determine the position of node $\mathbf{n}_e$ in subgraph R_3. In some cases, available information may be insufficient for node placement. For instance, node $\mathbf{n}_e$ in subgraph R_4 in Figure 4.29 revolves around $\mathbf{AP}_D$, although it can be assumed that its distance to other nearby nodes exceeds the reachability radius. Otherwise, a connection may exist. Other ambiguities may occur. However, their impact is limited by considering relative distance to other nodes in the network. Relational information will grow as the number of link connections increases [24].

During the relational location estimation process, further impairments may occur, especially those related to the channel since in practice the propagation environment is subject to multipath, shadowing, and scattering phenomena that induce distance estimation errors. Thus, estimated internode distance $\delta_{i,j}$ becomes $\delta_{i,j} = d_{i,j} + \Delta$, where $d_{i,j}$ is the true distance and Δ is a random variable whose characteristics are environment dependent. This random variable can be considered as a proportion of the true distance $d_{i,j}$.

Assuming that the random variable Δ has a mean value $E(\Delta) = \xi d_{i,j} < \infty$, which is a proportion ξ of the true distance $d_{i,j}$, we can write the estimated distance between nodes $\mathbf{n}_i$ and $\mathbf{n}_j$ as $\delta_{i,j} = d_{i,j} + \eta d_{i,j}$, where η is a non-negative random variable with mean $E(\eta) = \xi$. Using the minimization of the mean square error, we can see that in this case the effect of measurement noise can be reduced by scaling down the estimated distance by a factor $\frac{1}{1+\alpha}$, where α is a non-negative constant chosen to minimize a mean square error criterion, as explained in the following paragraphs.

The distance estimation is scaled by the use of a non-negative constant chosen to minimize the mean square error criterion. We follow the same notation, but the subindices are dropped for ease of presentation. Let $\delta = d + \Delta$ be a noise-corrupted distance measurement where d stands for the actual distance and Δ is a random variable such that its mean value is given by $E[\Delta] = \xi D$; δ can then be modeled as $\delta = d + \eta d$, where η is a random variable with mean value $E(\eta) = \xi$. Since δ is an available measurement, d could be obtained as $d = \frac{\delta}{1+\eta}$; and since η is unknown, we propose to estimate d as $d^* = \frac{\delta}{1+\alpha}$, where α is chosen to minimize the mean square error $E\{(d - d^*)^2\}$. This formulation leads to

$$E\left\{(D - D^*)^2\right\} = E\left\{\left(\frac{\delta}{1+\eta} - \frac{\delta}{1+\alpha}\right)^2\right\}. \tag{4.26}$$

Taking the derivative with respect to α, and making it equal to zero, we obtain

$$\frac{d}{d\alpha} E\left\{\left(\frac{\delta}{1+\eta} - \frac{\delta}{1+\alpha}\right)^2\right\} = E\left\{\frac{1}{1+\eta} - \frac{1}{1+\alpha}\right\} = 0, \tag{4.27}$$

which we use to find the value of α as

$$\alpha = E\left\{(1+\eta)^{-1}\right\}^{-1} - 1. \tag{4.28}$$

In the case of η being exponential,

$$E\left\{(1+\eta)^{-1}\right\} = \int_0^\infty \frac{1}{1+\eta} \frac{1}{\xi} e^{\frac{1}{\xi}\eta} = \frac{1}{\xi} e^{\frac{1}{\xi}} E_1\left(\frac{1}{\xi}\right), \tag{4.29}$$

which can be numerically approximated as (see [19, 25])

$$E\left\{(1+\eta)^{-1}\right\} \frac{1 + a_1\xi + a_2\xi^2}{1 + b_1\xi + b_2\xi^2}. \tag{4.30}$$

where $a_1 = 2.334733; a_2 = 0.250621; b_1 = 3.330657;$ and $b_2 = 1.681534$.

In order to assess the performance of the relational location estimation process, distance errors between the obtained locations and the true locations are computed and averaged for all points in the network. In other words, if the final position obtained by the estimation for node $\mathbf{n}_i$ is (x_i^*, y_i^*) and the true position is (x_i, y_i), we use the Euclidean distance to obtain the error measurement for each node $\mathbf{n}_i$ in the network. That is,

$$\varepsilon = \sqrt{(x_1 - x_i^*)^2 + (y_i - y_i^*)^2}. \tag{4.31}$$

Then we average all the nodes in the network to obtain, for a network with N nodes, a mean location error of

$$\varepsilon = \frac{1}{N} \sum_{i=1}^N \varepsilon_i. \tag{4.32}$$

Reported results, in this section, correspond to random generated networks, where the positions of the nodes are outcomes of a Poisson planar process and the network connection links were generated with a Dijkstra algorithm under a reachability radius–constraint assumption. Monte Carlo simulations were conducted for 5000 realizations of the network while maintaining the position of all APs constant. In order to obtain general results both mean location error and mean estimation error are normalized to the longest distance between two APs

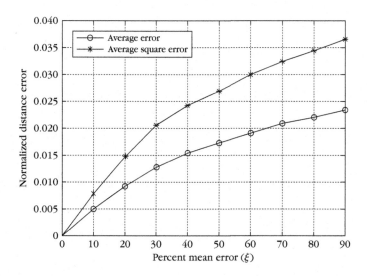

FIGURE 4.30

Normalized distance error for various propagation environments where η has been assumed exponentially distributed.

in the network since this distance does not change along the realizations. Also, this distance is related to the diameter of the network, and we assume that estimated distance errors are exponentially independent and identically distributed for all the links.

Figure 4.30 shows how the mean location error depends on the percentual mean estimation error ξ. We can see the location error increasing as the noise (ξ) in the environment increases. We also obtain the cumulative distribution of the realization error (see Figure 4.31). The figure shows the cdf for different noise-environment conditions ξ, and we can see, for example, with high probability (over 0.95) that distance error normalized is below 5% of the longest distance between two APs.

We observe the impact of the percentual mean error ξ on the average distance error in Figure 4.32, which shows the results for the networks with diverse node densities. We can see that for low densities, average distance error tends to remain below 0.2 (20% of the network diameter) for a wide variation of percentual mean error ξ.

Figure 4.33 represents the results of the location error of nodes for random networks with varying numbers of APs. We can see that error tends to diminish with an increase in the number of APs. Although the differences are minimal for low-noise environments (low values of ξ), the network is benefitted when more fixed known references are present in high-noise environments.

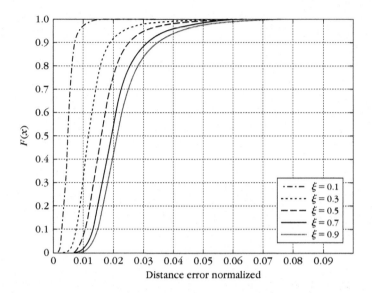

FIGURE 4.31

Normalized cumulative distribution of the average error.

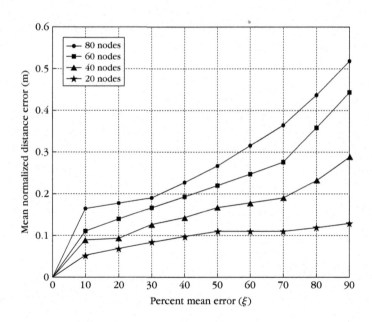

FIGURE 4.32

Normalized average position error.

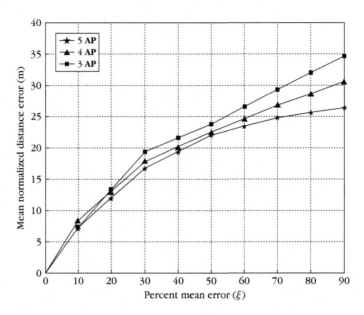

FIGURE 4.33

Normalized average position error as the number of APs varies.

A relational location method for PL acquisition in ad hoc and sensor networks has been introduced in this section. The proposed algorithm is based on hop length, nodes traversed by packets, relational neighbors, and statistical estimation. Basic shortest-path routing based on Dijkstra's algorithm was used to establish paths among APs, but the relational location estimation can be used with other routing algorithms. The algorithm was shown to work well even as node density increases. The accuracy of the algorithm was shown to be good for three, four, or five APs, resulting in no significant difference among these scenarios. The proposed algorithm is oriented to static environments. However, sequences of scenes allow obtaining mobility indices, both for a node on an individual basis and for a set of nodes on the network. The procedure could be used repeatedly to keep up with topology changes within certain time windows.

4.3 CONCLUSIONS

Sensors and nodes in ad hoc networks are not connected to immediate fixed references, which forces the development of alternative PL acquisition techniques subject to different uncertainties. Nevertheless, in some cases the proximity of region location may be sufficient for resource management or advertising services. In this chapter, several heuristic methodologies that provide proximity information were presented, including a few related single-hop scenarios and range-free PL acquisition, and others that focus on node separation (measured or in terms of the number of hops in a path). Relation or vicinity among the different nodes was also considered.

Simulation in various ad hoc environments was conducted, and results showed the feasibility of proposed heuristic techniques. It has also been shown that accuracy of the proposed schemes depends on routing spread characteristics and the quantizing error of the network area. Although the principal emphasis was focused on planar scenarios, some of the proposed techniques can be extended to 3D sensor environments.

REFERENCES

[1] D. Munoz, L. Suarez, R. Cruz, Position location scheme for low coverage areas, IEEE Mobile Communication Technology Conference (2004) 291-295.

[2] D. Niculescu, B. Nath, Ad hoc positioning system (APS) using AOA, in: Proceedings IEEE Infocom, 3 (2003) 1734-1743.

[3] C. Castro, Multidimensional scaling applied in position location systems, M.Sc. Thesis, ITESM, 2002.

[4] A. Abdi, H. Hashemi, S. Nader-Esfahani, On the PDF of the sum of random vectors, IEEE Transactions on Communications, 48 (1) (2000) 7-12.

[5] P. Beckmann, Probability in communication engineering, Harcourt Brace & World, Inc., 1967.

[6] O. Hernandez, Dead reckoning in ad hoc networks, M.Sc. Thesis, Monterrey Technology, 2006.

[7] N. Patwari, A.O. Hero III, Location estimation in wireless sensor networks, 36th Asilomar Conference on Signals, Systems and Computers, 2 (2002) 1523-1527.

[8] J. Mauve, J. Widmer, H. Hartenstein, A survey on position based routing in mobile ad hoc networks, IEEE Networks Magazine, November/December 2001.

[9] R. Torres, Técnicas de localización heurísticas en redes ad-hoc y de sensores, M.Sc. Thesis, Monterrey Technology, 2009.

[10] L. Leithold, The Calculus, 7th edition, Oxford University Press, 1994.

[11] P.G. Hoel, S.C. Port, C.J. Stone, Introduction to Probability Theory, Houghton Mifflin, 1971.

[12] M.D. Yacoub, Foundations of Mobile Radio Engineering, CRC Press, 1993.

[13] J. Gudmundsson, C. Levcopoulos, G. Narasimhan, Approximation minimum Manhattan networks, Nordic Journal of Computing, 8 (2) (2001) 219-232.

[14] R. Villalpando Hernández, Heuristic methods based on Manhattan model for position location in ad hoc networks, M.Sc. Thesis, Monterrey Technology, 2003.

[15] P. Prasitsangaree, P. Krishnamurthy, P.K. Chrysanthis, On indoor position location with wireless LANS, 13th IEEE International Symposium on Personal, Indoor, and Mobile Radio Communications, 2002.

[16] A. Savvides, H. Park, M.B. Srivastava, The bits and flops of the N-hop multilateration primitive for node location problems, Mobicom Workshop on Wireless Sensor Networks and Application, September (2002) 112-121.

[17] C.M. Hamdi, J.P. Hubaux, GPS-free positioning in mobile ad hoc networks, Journal of Cluster Computing, April 2002.

[18] R. Flickenger, Building Wireless Community Networks, 2nd edition, O'Reilly, 2003.

[19] E. Sanchez, Localización de nodos en redes de sensores utilizando relaciones de proximidad, M.Sc. Thesis, Monterrey Technology, 2006.

[20] A. Rao, S. Ratnasamy, C. Papadimitriou, S. Shenker, I. Stoica, Geographic routing without location information, ACM Mobicom, September (2003) 96-108.

[21] A. Tanenbaum, Computer networks, 4th edition, Prentice Hall, 2002.

[22] R. Yamamoto, H. Matsutani, H. Matsuki, Position location technologies using signal strength in cellular systems, in: Proceeding of the 53rd IEEE Vehicular Technology Conference, May 2001.

[23] J. Caffery, G.L. Stuber, Overview of radiolocation CDMA cellular systems, IEEE Communication Magazine, 36 (4) (1998) 38-45.

[24] L. Guibas, Sensing, tracking, and reasoning with relations, IEEE Signal Processing Magazine, 19 (2) (2002) 73-85.

[25] M. Abramowitz, I.A. Stegun, Handbook of Mathematical Functions, Dover Publications, Inc., 1972.

Terrestrial-Based Location Systems

Once the position location (PL) problem has been described, we review the fundamentals of wireless networks to develop a thorough understanding of the basic functions involved in the scenarios for position location. PL will be seen as part of the wireless network system, and mobility needs to be discussed because it imposes limits on network performance. It is also important to introduce how these scenarios evolve according to technology within new paradigms and architectures such as cooperative networks, cognitive radio, multihop scenarios, and reconfigurable networks.

Methods to obtain PL can be divided in two groups. First are the mobile-based methods where the mobile unit estimates its location based on signals received from base stations, for example, or from the global positioning system (GPS). Second are the network-based methods where the location of mobile stations is estimated by the processing of signals received at the base stations. To estimate location, the measurement of signal parameters (e.g., AOA, TOA, DOA, TDOA) is carried out by the base station (BS) and then this information is processed to estimate the position. In previous chapters, several techniques and their characterization have been introduced. In this chapter, we discuss the basic evolution of wireless networks and how position location could be considered from the networking point of view.

5.1 FROM CELLULAR TO RECONFIGURABLE NETWORKS

Wireless communication systems have evolved during the last years at an incredible pace—from simple analog techniques (first generation [1G]) to digital wireless systems (second generation [2G]) for personal use, and during this decade to some multimedia communication through third-generation [3G] wireless systems, as well as the sharing of resources having communication links traversing different networks. This section highlights the evolution, developments, and trends of wireless network technologies.

Wireless networks have evolved from several inventions throughout history, but one of the most important developments that form the bases of the systems that we know and use these days is the cellular concept. In 1933, the Federal Communications Commission (FCC) authorized the use of four channels in the 30- to 40-MHz range. During World War II, spread spectrum was used. In 1945, AT&T Bell Labs began experiments with the use of higher frequencies with the goal of improving mobile services and of introducing the cellular concept, where the network service area is divided into smaller regions called *cells* and each one of these cells is served by a low-power transmitter. In 1963, AT&T Bell Labs demonstrated the first cellular system.

In 1972, Motorola demonstrated the cellular telephone handset to the FCC. In 1974, the 800- to 900-MHz portion of the UHF band was allocated for cellular use. In 1983, the first commercial cellular system became operational in Chicago, the start of the Advanced Mobile Phone System (AMPS). Nordic Mobile Telephone Service (NMTS) in 1981 and Total Access Communication System (TACS) in 1985 became operational, in Scandinavia and the United Kingdom, respectively. In 1991, technology was introduced as well as the first validation of the U.S. Time Division Multiple Access (TDMA) standard. Also in 1991, the initial validation of Code Division Multiple Access (CDMA) started. Groupe Speciale Mobile (GSM) was introduced as the Pan-European Cellular System in 1992, and in 1994 cellular digital packet data systems were introduced. In 1994, the first personal communications services calls were placed (TDMA and CDMA).

The concept of personal communication systems comes to mind whenever we talk about a wireless network, so that the paradigm of point-to-point communication becomes person-to-person anywhere communication. Traditional networks were designed for a single service, voice. When information theory started gaining fame, people at remote sites needed to share information and so networks had to provide another service, data transmission. Demand for new services such as image retrieval and video transmission increased and with a little patience one can get these services with acceptable quality. The new era of communication networks comprises all such services at high-quality levels and for large numbers of users with small portable and mobile communication devices. Modern communications systems exploit traditional communication technologies such as twisted pair and coaxial cable, and integrate new technologies such as fiber and wireless links (cellular, satellite, etc.), which in turn provide mobility to users to achieve a true person-to-person anywhere communication.

The development of wireless communications has determined industry technology standards that depend on multiple access techniques of transmission media, the frequency band used, type of modulation, channel bandwidth, and type of service provided. The techniques used are Frequency Division Multiple Access (FDMA), TDMA, and CDMA. In FDMA, a channel of those available from the frequency band is occupied by the user. In TDMA, time is divided into slots and each customer uses one slot to communicate in a channel; the same channel can be used by other customers as long as they communicate in another time slot. In CDMA, all customers use the same channel, but they are identified by a code that is unique for each user and that does not interfere with those of others.

The modulation techniques most widely used are based on type of system. For example, in an analog system FM is used; for digital systems, several modulation schemes have been used: quadrature phase shift keying (QPSK), frequency shift keying (FSK), Gaussian minimum shift keying (GMSK), $\beta/4$ differential PSK (DPSK), binary PSK (BPSK), quadrature amplitude modulation (QAM), and so on. Most of these are treated in digital communication books (e.g., [21, 32, 67, 82]. The services provided are basically cellular, personal communication systems (PCSs), paging, cordless, and specialized mobile radio (SMR).

The first U.S. cellular telephone system was deployed by Ameritech in Chicago in late 1983. It used the standard advanced mobile phone system (AMPS), for which the FCC had just allocated 40 MHz of spectrum in the 800-MHz band. Each channel, one-way, has a bandwidth of 30 kHz, so the duplex channel has a total bandwidth of 60 kHz; therefore, the total 40 MHz contain 666 duplex channels. The modulation technique used is FM with a multiple access scheme of FDMA. Hence, this is an analog system. The channels in the 800-MHz band are numbered from 1 to 1023, excluding the numbers 800 to 989. If the reverse channel uses channel n, from the reverse channel band, the forward channel will use the nth channel from the forward channel band. The forward and reverse channel center frequencies are separated by 45 MHz.

5.1.1 Cellular Network Scenario

The cellular scenario is based fundamentally on the cellular concept, where an area is divided into cells with antennas transmitting at a lower power. This scenario has at its heart the concept of using the same carrier frequency in different areas or regions separated by distances such that the co-channel interference does not keep them from using the same frequency in those regions. Frequency reuse has been common for many years; we find it in the AM/FM radio broadcast stations around each country where the same frequency band can be used in two distant cities for two different radio stations as long as their signal satisfies the co-channel interference criterion.

The basic cellular scenario consists of these areas or cells being serviced by low power antennas and coordinated by a switching center. Figure 5.1 shows

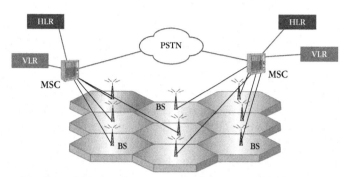

MSC: Mobile switching center; BS: Base station; PSTN: Public-switched telephone
network; HLR: Home location register; VLR: Visited location register

FIGURE 5.1

Basic cellular network scenario.

the basic architecture of a cellular network with hexagonal cells. It contains a
mobile switching center (MSC) with connectivity to the public switched telephone network (PSTN). Several cells can be connected to an MSC in high-density
areas where the number of MSCs needed increases. The MSC is in charge of
functions such as mobile registration, mobile paging, call establishment, handoff management, channel and resource assignment, signaling, connectivity, and
so on. It also has two databases that contain information about current users registered in the network: the home location register (HLR) and the visitor location
register (VLR) that register customers whose service contract is with the home
network or those that are roaming in this area, respectively. The information on
each user with a contract in an area is registered in the HLR, which contains
information such as the type of services to be provided.

The network is basically divided into regions called location areas; these areas
are cells or groups of cells where users register in the network. Thus, the HLR
and VLR registers contain user information such as the electronic serial number
(ESN), the mobile identification number (MIN), the user's profile, service and
restrictions, the user location given by the location area, and the base station
servicing the user. The MSC together with the HLR or VLR stores the location
area of the user. If a mobile station crosses the boundary between two different
location areas, an update is initiated by the network MSC to register the new
location area. This updating can be done periodically by the network regardless
of location area changes.

The first generation of cellular communication networks was provided
mainly by the standard AMPS among others. A brief review of some of these standards and their basic characteristics appears in Table 5.1. As shown in the table,
we can see common denominators, such as the FM modulation and the medium

Table 5.1   First-Generation Characteristics of Cellular Standards

Numerical Analysis Parameters						
	AMPS	**NAMPS**	**ETACS**	**NMT-900**	**JTACS**	**NTACS**
Year of introduction	1983	1992	1985	1986	1988	1993
MAC	FDMA	FDMA	FDMA	FDMA	FDMA	FDMA
Band (MHz)	824–894	824–894	900	890–960	860–925	843–925
Modulation	FM	FM	FM	FM	FM	FM
Channel (KHz)	30	10	25	12.5	25	12.5

access control (MAC), determined by FDMA. In addition, we see that depending on regions of the world, the frequency bands where the systems work vary slightly.

5.1.2 2G and 3G Technology Review

When one thinks about the evolution from the first generation to the third generation (3G), one can describe such technologies by a key characteristic. For example, first generation was characterized by analog modulation, while 2G is typified by the idea of digitalizing the transmission—that is, the use of digital modulation, which requires information to be digitized. 3G has been characterized by the term *multimedia*, although one may find that it has some limitations in order to achieve quality of service for each application (Figure 5.2). One of the most important facts of this evolution is that it has provided the base for incredible growth in terms of user numbers, unimaginable at the beginning of the 2G era.

The evolution of wireless personal communications was determined basically by the solutions to two problems. First, a new technology that did not have the same problems as the 1G system in terms of roaming and area code numbers was designed in Europe. Due to interests in evolving technologies with more potential growth, digital technologies were the principal option. The solution in this case was GSM. Second, when bandwidth demand in the United States was higher than what service providers could offer, the FCC granted more bandwidth in exchange for a long-term solution. Digital technologies were considered due to their evolutionary potential, and the U.S. TDMA and CDMA were the technologies chosen.

The fundamental concepts defining each digital technology are the modulation and MAC method used. There are several options, but only a few of these concepts were suitable for integrating the circuits in portable and mobile handsets.

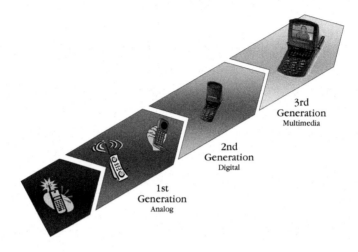

FIGURE 5.2

Evolution toward 3G.

Table 5.2 Second-Generation Characteristics of Cellular Standards				
Numerical Analysis Parameters				
	IS-95	**DCS-1900**	**GSM**	**PDC**
Year of Introduction	1993	1994	1990	1993
MAC	CDMA	TDMA	TDMA	TDMA
Band (MHz)	824–894, 1800–2000	1850–1990	890–960	810–1501
Modulation	QPSK	GMSK	GMSK	$\pi/4-$DQPSK
Channel (KHz)	1250	200	200	25

Table 5.2 presents a summary of the modulation, channel bandwidth, and multiple access used by 2G technologies. The main standard for 2G is GSM, especially in Europe. In America, GSM1900 (a variation of GSM) and CDMA (IS-95) are both used. One of the limitations of these technologies is data rate and capabilities of handling data applications such as Web browsing. For example, in the beginning GSM and CDMA could only offer data rates of up to 9.6 Kbps.

Figure 5.3 shows a visualization of the concept of FDMA compared to TDMA; in the latter, the time dimension was added to the frequency dimension already used in FDMA. Figure 5.4 displays CDMA, which adds the code dimension to the frequency.

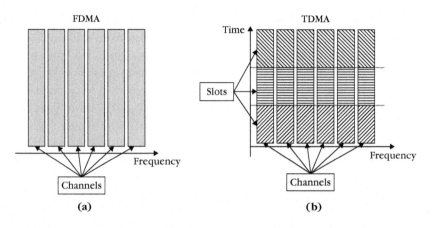

FIGURE 5.3

FDMA and TDMA concepts. (a) One user per channel. (b) One user per channel per slot.

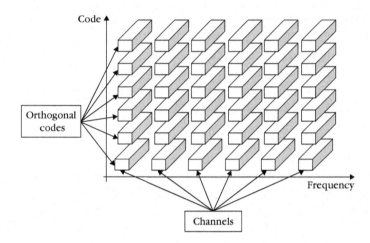

FIGURE 5.4

CDMA concept with N codes and N users per channel.

The most important parts of CDMA that explain its popularity are power control and orthogonal codes. Even though signals from users are not necessarily synchronized, the codes have properties of quasizero cross-correlation that allow them to eliminate multiple access interference (MAI). Power control is necessary in order to extract the information from the superposition of the signals from all the users transmitting, since it allows a fair comparison of these signals by not letting any of them be stronger to the extent of eliminating the desired information.

Since increased demand for service and for new applications is a market constant, 2G systems needed to evolve. Such upgrades were termed 2.5G. The enhanced systems have as many features as their 3G evolution has. The enhancements of 2G toward 2.5G were concentrated in the area of data services; technologies such as general packet radio services (GPRS) became available. Other technologies included enhanced data rates for global evolution (EDGE), originally defined as enhanced data rates for GSM evolution, and high-speed circuit-switched data (HSCSD). HSCSD is the system that allows handling the frequency channels the same as in a normal GSM system, but it lets the network assign several time slots to the same user, hence increasing data rate. The upgrade did not require hardware changes in the network, only new mobile handsets that could use HSCSD. The problem with this technology is that frequency and time are used in a circuit-switched fashion, which makes some resources unavailable to other terminals even if the resources are not being used at specific times. A point in favor of HSCSD is that since information and resources are handled according to a circuit-switched criterion, it works with real-time services. In contrast, GPRS, which is the most used, is not well suited for real-time applications. When GSM is upgraded with GPRS, the eight time slots are used and the data rate increases up to 115 Kbps. This is achieved by using a packet-switching service, which allows the occupancy of the resource by one user only when there is a packet to be transmitted. The most important applications in GPRS are email and Web browsing, which do not require real-time delivery of information.

GSM uses a modulation known as GMSK where information symbols are processed by a Gaussian filter before modulation. EDGE, in contrast, changes the modulation scheme to 8PSK, which allows transmission of three bits per symbol in the same bandwidth; thus, the data rate is three times higher than for GSM. EDGE can coexist with GSM, allowing users to keep their handsets when not requiring the higher data rate services. Due to distortion caused in the channel and path loss, 8PSK is not suitable for long distance; hence, the combination of data transmission systems of GPRS and EDGE, known as enhanced GPRS or EGPRS, can be deployed to service data and voice. The maximum data rate achieved with EGPRS using the eight time slots of the frequency channel is 384 Kbps, which is the standard data rate handled in 3G systems.

The standard IS-95 CDMA started providing a data rate of 9.6 Kbps, and later incremented to 14.4 Kbps in the first version. The standard was reviewed and in its B version incremented its data rate to 64 Kbps. Version IS-95C works with data rates of up to 144 Kbps. This last version represents a smooth transition to 3G toward the standard cdma2000; in brief, there is a direct evolution of IS-95 to 3G with cdma2000, but not in the case of GSM, which needs to evolve toward WCDMA. The set of protocols together with the standard IS-95 is known as cdmaOne. The power control mechanism adjusts the transmission power in 1-dB increments 800 times per second.

For the 3G standards, complete consensus did no exist in terms of technologies, and two standards were mainly established, both based on CDMA. Although there are other proposals for 3G such as OFDM and a hybrid version of TDMA and CDMA, the cdma2000 by the Telecommunications Industry Association (TIA) and the wideband CDMA (WCDMA) by the European Telecommunications Standard Institute (ETSI) are the most used at this moment.

The wideband CDMA bandwidth is 5 MHz to accommodate data rates such as 144 Kbps and 384 Kbps, and up to 2 Mbps with favorable conditions on wireless transmission. Another important characteristic is the ability to work with different data rates on a frame-by-frame basis by using the variable spreading factor, which is the ratio of the transmission bandwidth occupied and the input bit rate. An improvement of the traditional power control of cdmaOne was also included to work with increments of up to 1 dBm and faster on both the forward and reverse channels. This standard is the evolution of GSM/GPRS, providing all services from the first day of operation. WCDMA is not backward compatible with cdmaOne. The specifications of WCDMA are developed by the 3G Partnership Project (3GPP) consortium.

cdma2000 is the 3G standard from TIA, which is backward compatible with cdmaOne, and its specifications were developed by the 3G Partnership Project 2 (3GPP2) consortium. IS-2000, known as cdma200, has also evolved. At the beginning in 1999, it had a data rate of 316.8 Kbps; version A in 2000 included voice and data. In 2002, version C was developed to have a data rate of 3.1 Mbps for the forward link, and in 2004 version D had a reverse link with 1.8 Mbps. Also in 2004, IS-856-A was developed to enhance the 3.1 Mbps of the forward link.

3G technologies provide various services such as telephony, short messaging, emergency calls, Internet access, packet and circuit switching, and more. In terms of applications, earlier technologies work in 3G (e.g., the Wireless Application Protocol (WAP), Java, Bluetooth, Wireless Markup Language (WML), and so on). Applications such as voice, messaging, Internet access, and the current location-based services (LBS) are also available.

LBS determines the location of a customer and the information is used to provide services according to location. For example, a user's location information can be employed in a map to help find the way to a destination. In this case, location should be accurate in order to position the user on the map. Location of users is also important to service providers to develop various network areas with different service charges or a dynamic charging system that changes with time according to events. For example, user PL is important in determining channel allocation strategies in base stations (BSs) where attractors are established, since at certain times the attractors might represent traffic congestion spots.

It is also important that location information of a mobile provides marketing data to many organizations and also to the user trying to locate a specific type of business. Applications such as tracking can also be implemented as an LBS; only a handset is required to track a vehicle.

Regarding user location, the network has the user's location registration made by the MSC and the HLR or VLR, which only provide information on the location area and the BS at which the user is found. This simple method is based only on the network coverage area and could be more accurate by sectoring the cells. In general, a method such as the TDOA discussed in Chapters 2 and 3 can be used, where measurements of signal parameters form hyperbolas like those shown in Figure 5.5, and after computing geometric relationships of the hyperbolas, an estimate of the position can be obtained.

The basic architecture for any network is presented in Figure 5.6, where the signal from the mobile is received through the radio access network (RAN), and depending on the service (voice or data), a location center (LC) can determine location with the help of the MSC, HLR, or VLR using algorithms described in previous chapters.

Talking specifically about GSM, we must recall that it is a system based on TDMA with eight time slots per frequency channel, occupying a transmission

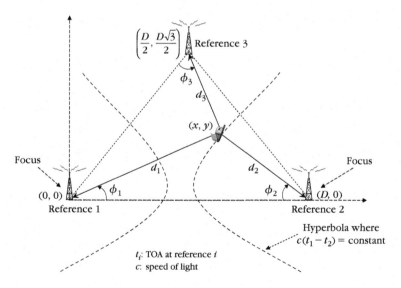

FIGURE 5.5

TDOA position-location technique.

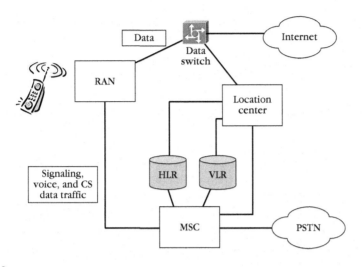

FIGURE 5.6

Basic location architecture.

bandwidth of 200 KHz. The time slots are grouped into frames and then in multiframes. Each of the multiframes carries logical channels that transmit control signals or information traffic. Within one of the time slots in a frame, a 26-bit training pseudorandom sequence can be used to estimate the channel by performing the autocorrelation of the received sequence and a locally generated one. This autocorrelation can be used in localization by considering that the maximum value of such autocorrelation indicates time reference. The base stations also transmit a synchronization sequence in the synchronization logical channel. This sequence can also be used to estimate time reference by computation of the autocorrelation. Due to the use of GMSK modulation, the autocorrelation peak is approximately 4 bits wide, which is an accuracy limitation for a time-based method. This inaccuracy is worsened by the multipath effects of the channel. If the receiver is not prepared to mitigate these effects, the peak of the autocorrelation would be shifted from its original place depending on the effects over the amplitude, phase, and time delay of the received signal. Signal propagation time could be estimated using the time advance (TA) feature used by the base station to indicate to the mobiles the amount of time that the transmission must advance in order to receive the mobile's signal in the corresponding time slot. TDOA can be estimated using the observed time difference (OTD) that is used to help the handover mechanism where each mobile unit monitors the base stations around it. AOA, although not very accurate, could be estimated using cell sectorization. All of these techniques help to determine the circles, hyperbolas, and so on, that could be used geometrically to estimate PL as discussed in Chapters 2 and 3.

The implementation of PL in GSM could be done considering mobile-based or network-based positioning. In mobile-based positioning, the mobile handset must be equipped so that it can receive signals from at least three base stations and compute TDOA to self-determine its position. In order to do this, the base stations must be modified so that the signals from the three base stations transmitted to the mobile station are synchronized. The use of the broadcast control channel (BCCH) is appropriate since no power control or frequency hopping is used that could affect accuracy of estimation.

For the network-based PL, we have the option of three base stations monitoring the mobile signal through the traffic channel (TCH). These measurements, which are basically of TOA, are shared in a system generally called the location service center (LSC), which computes the TDOA parameters and performs the estimation with methods discussed in previous chapters. The LSC system has direct communication with the MSC and to the HLR and VLR in order to use the information to estimate the PL.

In a CDMA system, the reverse traffic channel is active only when a call is active, making this channel practically useless for PL unless a service such as emergency E911 is requested via a call. The other option is to use the reverse access channel where a mobile answers to commands sent by the base station through the forward channel. The registration update of the mobiles can be used as a signal to estimate PL. This registration update can be established periodically through a command by the base station, which will have an effect on the signaling load of the base station.

The code acquisition and tracking used in the CDMA receivers can be helpful for estimating time delay [31] through the use of sequence correlation. The estimate would have an error based on a chip duration. After that, a system such as the LSC could be used to compute the TDOA from several measurements related to the various base stations. One of the limitations for location algorithms in a CDMA network is the MAI caused by other active users, not only on the same base station but also from other cells. With the use of power control, the signal of the mobile to be located arrives at its designated base station with no strength problem because all other active users have power control as well. On the other side, the signal from the mobile to be located will not arrive at the other base stations involved in the location algorithm with the same strength as if the mobile were in such a cell; all other users designated to such a base station will contribute to the MAI caused by the signal used for location. This is a classic problem of the near–far effect. For emergency calls, the mobile could be programmed to transmit at maximum power regardless of distance to its servicing base station. This would mitigate the near–far problem in the other base stations involved. Solutions that use multiuser detectors or interference mitigation techniques to estimate TOA or TDOA have been proposed in the literature.

5.1.3 4G and Beyond

The evolution of personal wireless communications characterizes 4G, where the main objective is to use various service platforms and access technologies in a unified system that provides services in an existing infrastructure. This integration of service platforms may be able to provide alternative wireless operators with a chance to compete with 3G systems. 4G has also been known as broadband 3G (B3G).

4G systems' aim is to provide high network capacity and broadband connectivity so that handsets are able to receive high-definition video; universal coverage so that the anywhere-anytime and any-technology paradigm can be fulfilled; and the ability to form personal networks at low cost.

The integration of service platforms and technologies challenges network operators to provide a seamless mobility among platforms and technologies. Technologically speaking, 4G systems are expected to rely on turbo coding and space–time coding techniques, a modulation based on QAM with OFDM (i.e., multicarrier QAM), and multiple access still to be defined that integrates information and coordination of interference.

The future 4G networks will incorporate 2G and 3G systems and a wide variety of networks with heterogeneous devices that will be characterized by smart communications techniques. It Integration is expected to include WCDMA, GSM, cdma2000, wireless local area networks (WLANs), satellite networks, digital video broadcasting (DVB), digital audio broadcasting (DAB), and newly emerging networks. So instead of having a single network do everything, a variety of networks with different kinds of access will provide all types of services. These networks will have to cooperate so that access to services and applications is transparent to the user.

The smart devices will be able to download configuration files from a service provider to adapt to requirements for accessing a given application. This procedure will make such devices reconfigurable. The 4G network will likely be organized in layers (not hierarchical ones necessarily) where devices can communicate among themselves (e.g., using Bluetooth or WLAN technologies in a peer-to-peer fashion). Also, access points (APs) will comprise a layer providing services such as Internet access. Cellular 2G and 3G networks will work as transport networks providing wireless access to all WLAN APs in order to reach final destinations or universal connectivity, (e.g., to a satellite network). Figure 5.7 shows an example of this layering network for 4G systems.

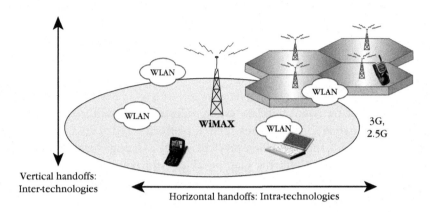

FIGURE 5.7

4G concept of integrating technologies.

One of the characteristics of this layered architecture is heterogeneous networks working collaboratively; to provide mobility, there should be consideration not only of internal handoffs, such as those occurring between adjacent cells of a network, but also of handoffs between different technologies. The latter have been called *vertical handoffs* due to the layered structure of the 4G paradigm. In Section 5.2, some discussion on this will be presented, but one must keep in mind that in order to perform such vertical handoffs, mobile handsets must be equipped with hardware that is capable of adapting to various radio interfaces. Energy consumption is affected by this feature since transmission power varies by technology. Thus, there is a trade-off between the use of different resources and the need to implement such adaptability.

Development of 4G architectures is concentrated on the flexibility at all levels of the architecture in order to improve performance not only in applications and services, but also in air interface and network control and signaling.

5.1.4 Ad Hoc and Sensor Network Scenarios

This section introduces a general overview of wireless mobile ad hoc networks (MANETs) and sensor networks, and their evolution, applications, and selected issues such as connectivity and scalability. This section is intended to prepare the reader for a subsequent scenario.

A MANET (see examples in Figure 5.8) is a dynamic network formed by a collection of arbitrarily located wireless mobile nodes without the use of existing network infrastructure or centralized administration. MANETs begin with at least two nodes broadcasting their presence (beaconing) with respective address information; a mobile node communicates with another directly via a single hop whenever a radio channel with adequate propagation characteristics

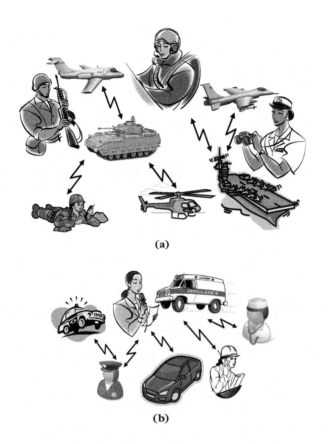

FIGURE 5.8

Two examples of ad hoc networks.

is available between them. If devices are not directly connected, multihop communication is necessary where one or more intermediate nodes must act as a sequence of relay nodes that forward the traffic of the communicating nodes. As more nodes join the network or some of the existing nodes leave, the topology updates become more numerous and complex, and usually more frequent, thus diminishing the network resources available for exchanging user information [10, 12].

Wireless ad hoc networks have some characteristics similar to wireless networks, but the differences are what make them interesting to study. Some of these differences follow:

Mobility of nodes: An ad hoc wireless network does not depend on a fixed infrastructure. The network is reconfigurable following mobility and communication needs.

Point-to-point communication: In contrast to cellular networks where communication between two cellular devices is made throughout a base station, in ad hoc networks communication between nodes is point to point. This means that two nodes within their sensivity area can communicate directly and without a base station's help.

Self-organizing: An ad hoc wireless network is self-organizing and adaptive to the communication requirements of the moment. This is a fundamental feature of ad hoc networks. A terminal has to be capable of detecting other terminals that are within reach, and it also must learn which terminals are reachable through the network. Autoconfiguration also concerns detecting services available in the networks. Because of this characteristic, most of the protocols for traditional networks cannot easily be adapted to ad hoc networks.

Early MANET applications can be traced back to the DARPA PRNet project in 1972, in which a combination of Aloha and CSMA were used to support the dynamic sharing of the broadcast radio channel. SURAN was developed by DARPA in 1983 to address major issues in PRNet in scalability, security, and energy management. The design of low-cost packet radio (LPR) in 1987 featured a digitally controlled DSSS radio with an integrated Intel 8086 microprocessor-based packet switch. The Department of Defense initiated the DARPA GloMo Information Systems program in 1994, supporting Ethernet-type multimedia connectivity among wireless devices. WINGs at University of California–Santa Cruz deployed a flat peer-to-peer architecture, and the MMWN project from GTE Internetworking used a cluster-based architecture. Tactical Internet implemented by the U.S. Army in 1997 is the largest-scale deployment of a mobile wireless multihop packet radio network; DSSS and TDMA radio are used with data rates in the range of tens of kilobits per second. In 1999, ELB ACTD demonstrated the feasibility of Marine Corps war-fighting concepts that require over-the-horizon, beyond-LOS communications from ships at sea to Marines on land via an aerial relay. Lucent's WaveLAN and VRC-99A were used to build the access and backbone connections. In 1990, the IEEE, when developing IEEE 802.11, replaced the term *packet radio network* with *ad hoc network*. At present, the Bluetooth system is perhaps the most promising technology in the context of MANETs [12, 22], but there are problems with scalability, security, and coverage. These problems have been addressed and new technologies such as ZigBee have positioning capabilities.

A MANET can be used to exchange information between the nodes, allowing them to communicate with remote sites that otherwise would not be reachable or where no infrastructure (fixed or cellular) is available. Examples include tactical networks with related applications to improve battlefield communications or survivability, rescue operations, disaster recovery, collaborative work in remote areas, public events, construction sites, and so on [22]. Recently, the introduction of technologies such as Bluetooth and WiFi are helping to enable eventual commercial MANET deployments outside the military domain [10, 12]. Futuristic applications include home networks where devices can communicate directly to exchange information, encompassing computers, microwave ovens, door locks, and other appliances interconnected by a wireless network [22]. Sensor networks have been proposed for environmental monitoring, where the networks could be used to forecast water pollution or to provide early warning of an approaching tsunami. Perhaps the most far-reaching applications are autonomous networks of interconnected home robots that clean, perform security surveillance, and so on.

The multihop nature and the lack of fixed infrastructure add a number of issues and design constraints that are specific to ad hoc networking.

Connectivity issue. In MANETs, because nodes can move arbitrarily, the network topology can change frequently and unpredictably, resulting in connectivity changes, frequent network partitions, and possibly packet losses.

MAC issue. The MAC scheme is also difficult in ad hoc networks due to the time-varying network topology and the lack of centralized control; instead, carefully designed distributed MAC techniques must be used for channel resources [10, 30].

Scalability issue. Many MANET applications involve large networks with tens of thousands of nodes. Scalability is critical to the successful deployment of these networks, presenting many challenges in addressing, routing, location management, security, and so on [12].

Clustering issue. Efforts to maintain a relatively stable effective topology, limit far-reaching reactions to topology dynamics, rearrange clusters, and assign nodes indicate that establishing a good clustering algorithm can be complex, requiring excessive processing and communications overhead [47]. On the other hand, it is known that clustering is useful in improving MAC resource management and in stabilizing network topology [41] (See Lawson and Denison [43] for an excellent overview of this issue.)

All these challenges are potential sources of service impairment in MANETs, and hence may degrade the QoS seen by users of the network, [10].

Connectivity Issues in MANETs

A widely studied problem in the field of MANETs is the connectivity issue. A fundamental property of networks is that every node should be able to communicate with every other node in establishing connections, defining in this way a certain level of connectivity of the nodes, and at the same time the topology of the network. In a given network, such as the one depicted in Figure 5.9a, mobile node i may lose connectivity with the rest of the network simply because it has wandered off too far as in Figure 5.9b, or its power reserve has dropped below a critical threshold, or because of the influence of certain phenomena in the radio channel such as fading or shadowing. In these cases, coverage area is reduced and links with the other nodes of the network are likely affected, as in Figure 5.9c.

On the other hand, a mobile node in a network, such as node i shown in Figure 5.10a, can gain connectivity if it approaches close enough to the rest of the network as in Figure 5.10b, or because it increases its coverage area (as a result of a reload of its power reserve, or by the influence of constructive multipath effects), creating new links with the other nodes, as shown in Figure 5.10c [10].

In both cases, the connectivity of the nodes and therefore the topology of the network are affected. The analytical techniques contemplated for constructing the theoretical framework include the following methods:

Branching processes. Each individual in the network has a random number of children in the next generation in accordance with a certain probability distribution [38].

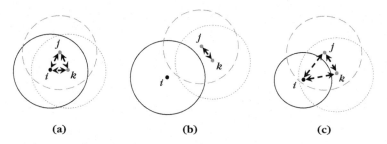

(a)   (b)   (c)

FIGURE 5.9

Connectivity losses of node i by separation and reduction of coverage.

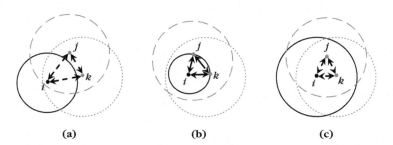

(a)   (b)   (c)

FIGURE 5.10

Connectivity gains of node i by approach and increment of coverage area.

Clustering and migration processes. These tell us about the formation of clusters in which individuals move between groups and can be studied by Markov processes [40, 43].

Random graphs. A graph consists of a set of vertices and edges; a graph is connected if and only if there exists a walk between any two vertices [28].

Point processes. These are stochastic processes where points are scattered over an *n*-dimensional space in some random manner [54].

Percolation theory. In this formulation, disks with equal radii are generated on an infinite plane according to a Poisson process; as the intensity of the process increases, disks overlap and form clusters. The theorem states that a finite critical intensity exists, above which a unique unbounded cluster almost certainly exists. A cluster of disks is equivalent to a connected network when the maximum transmission range is twice the disk radius. Percolation shows the farthest distance at which the broadcast of packets reaches nodes in the network, providing some long-distance multihop communication. We do not choose this method, because a received packet is not necessarily propagated since it has a specific destination address where routing techniques mark the path and the destination node [49].

Diffusions. A one-dimensional diffusion is a model of the motion of a particle with limited lifetime, continuous path, and no memory, traveling in a linear interval [36, 39].

All these options are worth pursuing as individual research entities. We selected the point processes method. Results and models for connectivity have been developed by Antonio [84].

Location in Reconfigurable Networks

The basic scenario for a PL application in reconfigurable networks can be seen in Figure 5.11. We have nodes with random coordinates in a rectangular area with certain connectivity defining a network topology. Some of those nodes are the APs that are in charge of connectivity to other networks or the Internet.

APs are primary pieces of a wireless network and they have a finite range. APs are actually the "hub" of a wireless network. Each AP is connected to a wired network and every node on the wireless network speaks to an AP, which connects that node to the rest of the world. These APs can have multiple links. Also, when nodes establish a node discovery and service discovery algorithm, APs will be the objectives in terms of applications, connectivity, IP addresses, servers, gateways, and so on.

An AP can be a location mobile unit (LMU). In global systems for mobile communications (GSM), several LMUs are introduced. LMUs have limited capabilities to support PL functions. We propose supporting the location process in ad hoc networks through the use of three APs in a cluster or in any area of the

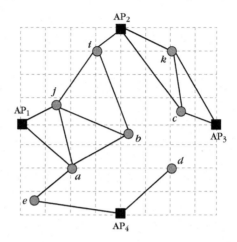

FIGURE 5.11

Fundamental scenario in reconfigurable networks.

ad hoc network. In most of the methods presented in Chapter 4 to solve the PL problem, three APs act as nodes in the network and provide fixed references in the network for PL purposes.

The evolution of network density and mobility makes necessary certain special nodes located at strategic points in the area so that connectivity is not compromised whenever node density is low. For example, at certain times during the day, there will be a few nodes in the area that might not be capable of establishing a multipath route to an AP. Those special nodes can then be turned on in order to provide coverage at those times every day. These special nodes can have features different from the regular nodes, such as higher transmission power to provide the needed coverage. Also, the nodes can have higher complexity and intelligence in order to turn themselves on and off according to network conditions, so they should be monitoring certain network and air interface variables.

The recent literature has reflected interest in location estimation algorithms for wireless sensor networks [55, 62]. Distributed location algorithms offer the promise of solving multiparameter optimization problems even with constrained resources at each sensor [70]. Devices can begin with local coordinate systems [7] and then successively refine their location estimates [1, 69]. Based on the shortest path from a device to distant reference devices, ranges can be estimated and then used to triangulate. Distributed algorithms must be carefully implemented to ensure convergence and to avoid error accumulation in which errors propagate serially in the network. Centralized algorithms can be implemented when the application permits deployment of a central processor to perform location estimation. In Celebi and Arslan [9], device locations are solved by convex optimization. Both Moses et al. [55] and Patwari et al. [62] provide ML estimators for sensor location estimation when observations are AOA, TOA [55], and RSS [62].

Since a classical multilateration process cannot be applied directly in an ad hoc environment due to the lack of direct connectivity of users to well-located APs, multihop algorithms are needed. The goal of typical multihop localization schemes is to estimate the position of all nodes in the network based on a few APs with known positions. Nodes in the proximity of APs are located first and then these nodes become new land references (with a certain degree of uncertainty) used to locate a new set of neighbors. This process continues in an iterative fashion until positions of all the nodes in the network have been estimated. This type of iterative algorithm suffers from error accumulation throughout the iterations and requires a considerable amount of processing from all nodes in the network. Many multihop localization algorithms such as APS [59, 60, 74] are geometric in nature and hence do not profit from statistical knowledge of the environment. Further, these methods require communication of the nodes with all immediate neighbors, and at some point they may even require the broadcasting of distance correction factors to the entire network, rendering them power inefficient.

In the literature, statistical multihop positioning schemes have been proposed in Savvides et al. [70], where the methods are based on accurate ranging measurements and linearized least-squares multilateration solutions. These methods require that each node with an unknown position be at a one-hop proximity from at least three land references (some may be APs, and some may be nodes that obtained position estimates from previous iterations of the positioning scheme). Further, the cited schemes rely on the solution of global nonlinear optimization problems to avoid error accumulation in the position estimates. Even when the computations are distributed through the nodes in the network, the amount of computational load required at each node may render these schemes impractical in many situations. Efforts to statistically characterize error-inducing parameters in multihop localization schemes have appeared in Savvides et al. [71] where ranging and AOA estimation errors are assumed to be Gaussian distributed.

We divide cooperative localization into centralized algorithms, which collect measurements at a central processor prior to calculation, and distributed algorithms, which require nodes to share information only with their neighbors, but possibly iteratively. Both methods are described in the following.

Centralized Algorithms

If the data are known to be well described by a particular statistical model (e.g., Gaussian or M-Erlang), then the ML estimator can be derived and implemented [63]. One reason that the ML estimators are used is that their variance asymptotically approaches the lower bound given by the Cramer-Rao bound (CRB). In this kind of estimator, the maximum of the likelihood function must be found. There are two difficulties with this approach.

1. *Local maxima*: Unless we initialize the ML estimator to a value close to the correct solution, it is possible that our maximization search may not find the global maxima.
2. *Model dependency*: If measurements deviate from the assumed model, the results are no longer guaranteed to be optimal.

One way to prevent local maxima is to formulate the localization as a convex optimization problem. In Doherty et al. [19], convex constraints are presented and can be used to require a node location estimate to be within radius r and/or angle range $[\alpha_1, \alpha_2]$ from a second node. Multidimensional scaling (MDS) algorithms [76] formulate sensor localization from range measurements as a least-squares (LS) problem. In classical MDS, the LS solution is found by eigendecomposition, which does not suffer from local maxima. To linearize the localization problem, the classical MDS formulation works with squared distance rather than distance itself, and the end result is very sensitive to range measurement errors.

Distributed Algorithms

Distributed algorithms have two advantages. First, for some applications, no central processor is available to handle the calculations. Second, when a large network must forward all measurement data to a single central processor, there is a communication bottleneck and higher energy drain at and near the central processor. Distributed algorithms for cooperative localization generally fall into one of two schemes.

Classical multilateration: Each sensor estimates its multihop range to the nearest land-reference nodes. These ranges can be estimated via the shortest path between the node and reference nodes, that is, proportional to the number of hops or the sum of measured ranges along the shortest path [59]. Note that the shortest path algorithm is already executed in a distributed manner across the network. When each node has multiple range estimates to known positions, its coordinates are calculated locally via multilateration [77].

Successive refinement: These algorithms try to find the optimum of a global cost function (e.g., LS, WLS, or ML). Each node estimates its location and then transmits that assertion to its neighbors [70]. Neighbors must then recalculate their location and transmit again, until convergence. A device starting without any coordinates can begin with its own local coordinate system and later merge it with neighboring coordinate systems [7]. Typically, better statistical performance is achieved by successive refinement compared to network multilateration, but convergence issues must be addressed.

Bayesian networks provide another distributed successive refinement method to estimate the probability density of sensor network parameters. These methods are particularly promising for position localization. Here each node stores a conditional density on its own coordinates based on its measurements and the conditional density of its neighbors. These algorithms are important given the lack of infrastructure that provides beacons or landmarks as references. A summary of several techniques, some of which have been described in this section, can be found in Basagni et al. [3].

5.2 MOBILITY IN WIRELESS NETWORKS

Mobility is a very important issue that needs to be addressed in a wireless network, since it limits capacity. For example, in a cellular network with no mobility, we can establish a capacity criterion based on the Erlang-B formula. Channels will be occupied by users that contribute to traffic so that a cell with C channels will have a blocking probability given by the formula. If we allow users to move, then some users from adjacent cells will hand off their calls to the cell of interest, producing a higher occupancy state in the cell and thus a higher blocking probability. This simple reasoning shows how mobility could limit capacity.

In the following sections, we introduce some aspects of mobility that need to be considered to balance the capacity–coverage trade-off that mobility causes.

5.2.1 Capacity and Coverage Issues

Demand for wireless services has increased, which in turn brings new issues to consider such as mobility management for service providers [13]. One of the major objectives of a telecommunications system is to offer a service of excellent quality based on user requirements. In order to evaluate how a network or a communication system performs, we need to define measures that quantify the effects of varying parameters such as demand and capacity. In wireless networks, we allow users to move from area to area, thereby causing handoffs in the network. A user requesting service for the first time from the network is considered a new call. Two of the most important performance measures in wireless networks are the new call and the handoff blocking probabilities [34, 48]. The handoffs are a fundamental feature in cellular systems; their performance and efficiency strongly depend on the use of adequate algorithms. For cellular communication systems, to ensure mobility and capacity, to maintain the desired coverage areas, and to avoid problems of interference, it is necessary to correctly assign the calls to the corresponding service areas in the entire cell and in the entire network.

In a CDMA network, soft handoff has been modeled considering overlapping areas, such as in Kwon and Suang [42], Miranda-Guardiola and Vargas-Rosales [53], and Scaglione et al. [72]. Admission of handoffs is done using one of three criteria. The first is to consider handoffs and new call arrivals equally for occupancy of the channels, the second reserves channels to give priority to handoffs, and the third sends them to a queue if no channel is available. Several performance evaluation algorithms have been introduced for these handoff strategies (e.g., see [34] and [79]) for reservation and queueing strategies and in McMillan [48] for the reservation and no-reservation strategies.

In a simplistic way, we can see each i-th cell as a resource with C_i channels offered Poisson traffic for new calls with λ_i calls per time unit, with exponential channel residence times with mean $1/\mu_i$. This would translate to considering each cell as an $M/M/C_i/C_i$ system with performance provided by the Erlang-B formula,

$$E\left(\frac{\lambda_i}{\mu_i}, C_i\right) = \frac{\left(\frac{\lambda_i}{\mu_i}\right)^{C_i} \frac{1}{C_i!}}{\sum_{k=0}^{C_i} \left(\frac{\lambda_i}{\mu_i}\right)^k \frac{1}{k!}}. \tag{5.1}$$

But we need to see that each cell receives offered traffic due to handoffs from adjacent cells, as shown in Figure 5.12. And since the success of the new call traffic depends directly on the available capacity of the cell to which it is being offered, and the handoff calls depend on the available capacity of the cell from which it comes, we can see a cellular network such as that in Figure 5.12 as a Jackson-type network of queues [5] as shown in Figure 5.13. In Vargas-Rosales

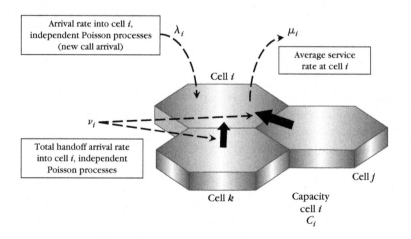

FIGURE 5.12

Traffic offered to cell in wireless network.

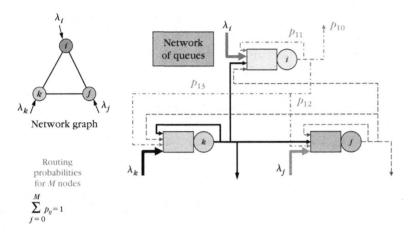

FIGURE 5.13

Cellular network as Jackson network of queues.

et al. [79], it was shown that this viewpoint has a solution for the blocking probability of new calls and handoff calls with channel reservation. The fundamental idea in this model is that the traffic offered to a cell is given by the new call traffic plus handoffs that have already been accepted in an adjacent cell; that is, if we denote as ν_{ij} the handoff traffic offered from cell i to cell j, B_i as the new call blocking in cell i, and B_{hi} as the handoff blocking in cell i, we can obtain the handoff rate out of cell j offered to cell i as

$$\nu_{ji} = \lambda_j \left(1 - B_j\right) p_{ji} + \left(1 - B_{hj}\right) p_{ji} \sum_{k \in A_j} \nu_{kj}, \tag{5.2}$$

where A_j is the set of neighboring cells to cell j, and p_{ij} is the routing probability from cell i to cell j. The first term in Equation (5.2) is due to those new calls offered to cell j that are accepted and after a residence time, handed off to cell i with probability p_{ji}. The second term is due to all handoff calls that are offered and accepted into cell j from its adjacent cells and then handoff to cell i.

One can see that this model helps us to comprehend the effects that mobility has on performance by varying the routing probabilities to increase the proportion of handoff calls being offered. To solve for the blocking probabilities, one needs to consider a fixed point because Equation (5.1) is in terms of traffic offered, and this depends on new call arrivals plus handoffs as given by Equation (5.2). Equation (5.1) now also has traffic from both arrivals of new calls and handoff calls. This immediately tells us that traffic increases, and thus blocking also increases; that is, capacity in terms of number of possible simultaneous users active is reduced. This model has been evaluated for FDMA and TDMA in Vargas-Rosales et al. [79], and for CDMA in Miranda-Guardiola and

Vargas-Rosales [53]. In addition, the use of reservation degrades performance for the type of traffic with higher bandwidth requests. Even though these networks have limitations in terms of interference levels, it has been shown that relevant limitations are due to blocking [37].

It is well known that CDMA capacity depends on the processing gain and the bit energy-to-noise ratio, but from another viewpoint, we can consider a scenario where a central cell is influenced by interference from an infinite number of rings (tiers), each of which contains cells transmitting at the same frequency with users working with perfect power control. The scenario was considered with homogeneous circular cells of radius R in Munoz et al. [56]. The advantages of using such a model are that we get bounds on the interference levels even for an infinite number of cells due to the infinite number of rings surrounding the center cell. In the model, the same number of users for each cell is considered, and in order to obtain the major influence of the users, in each cell all users are located at the closest point toward the center. The model used a simple propagation model with a path-loss exponent between 2.5 and 4, and cell radius was varied to consider cases with a radius of 3, 5, and 10 km. In the worst-case scenario, a cell capacity of 20 users was obtained when the number of interferents was infinity. Voice activity and sectoring were considered as well. The important aspect of this result is that regardless of the number of interferents, the capacity of CDMA cells with perfect power control will be lower-bounded by 20. We must be cautious when referring to this number since in these conditions FDMA and TDMA would be useless due to interference. The final result of the analysis in [56] is provided by the following lower bound:

$$N = \frac{(I/C) + 1}{1 + 8 \left(\frac{R+1}{2R}\right)^{\gamma} \left\{ \zeta \left(\gamma - 1, \frac{R+1}{2R}\right) + \left(\frac{R-1}{2R}\right) \zeta \left(\gamma, \frac{R+1}{2R}\right) \right\}}, \qquad (5.3)$$

where N is the number of users in each cell, γ is the path-loss exponent, R is the cell radius, C/I is the carrier-to-interference level usually set to -15 db, and $\zeta(x, y) = \sum_{m=0}^{\infty} 1/(m+y)^x$ is the Riemann-Hurwitz function that converges for $x > 1$ and $y > 0$.

In general, network capacity also depends on limitations encountered by the underlying channels. These limitations determine the data rates at which one can transmit with small bit error rates (BERs). Once the physical layer provides a reliable link to transmit, then the network functions take place, consuming some of the available bandwidth in order to achieve network control. So in order to consider capacity in wireless networks, we have to see that channel capacity or single-user system capacity and multiuser capacity need to be integrated. For networks with infrastructure, it is well known that the uplink will have a degraded performance once the number of users increases since interference will be an issue. The base station transmits at a certain power level in the downlink that is also affected by the amount of interference, creating a coverage problem in some areas since the downlink signal will not be received with as much power as it seems. We also know that higher frequencies will require higher sensibility from the receivers since received power is inversely proportional to frequency. For treatments of these capacity issues in single-user and multiuser systems, see Goldsmith [27].

For networks that have no infrastructure (i.e., reconfigurable networks such as ad hoc and sensor networks) capacity has been an important research issue. For these networks, it is not as simple since the concept of simultaneous number of users does not apply directly due to the distributed use of the bandwidth. In addition, issues such as bit rates, interference suppression, multiple access, geographic position, topology, connectivity, and reachability, among others, play an important role in determining the number of nodes that could be active at a given time in a network. Also, certain types of algorithms implemented could be improved if used in a distributed or cooperative way. Capacity in these networks has been studied in general [30], as has how mobility increases capacity when cooperation is used (e.g., see [29]). The work of Grossglauser and Tse [29] contains a study of a network with no mobility with nodes generated randomly on a disk or sphere, and as the node density increases, the throughput per origin destination pair decreases with a bound determined by $1/sqrt(n)$. It was also shown that this is the best performance that one can get even with optimal cooperation in relaying, routing, and scheduling. One issue would then be scalability, since the result in Gupta and Kumar [30] gives practically a zero throughput when the network grows. In contrast, mobility can help maintain constant origin–destination pair throughput even when the network grows, as shown in Grossglauser and Tse [29]. The result is based on the use of relaying as a form of multiuser diversity.

5.2.2 Modeling Mobility

In wireless networks, one needs to deal with the time variation of the topology due to mobility and the time variation of the channel due to interference, noise, environment, and other variables. The time variation of the channel imposes conditions on the establishment or breakage of wireless links, so that a physical layer issue has a relevant effect on the topology of networks such as those for reconfigurable devices. Diversity can help to overcome some of the problems presented at the physical layer due to the time variation of the channel, especially new diversity forms such as the cooperative kind where at some point there is always a device with a better channel response than other devices. In the end, performance experienced by users is the relevant objective that needs to be achieved by any operation or management function in the network. As discussed in the previous section, mobility imposes some restrictions on performance, where from the point of view of traffic we see that blocking is increased as mobility increases since handoffs increase.

The model introduced in the previous section provides a simple way to evaluate mobility in a cellular network by varying the proportion of handoff calls being offered through the use of routing probabilities in Equation (5.2). The model has been extended to the case of several classes of traffic by considering the solution of Markov chains for each of the cells and capturing mobility effects through the traffic equations. This has been shown in Vargas-Rosales et al. [79], in which a net revenue equation is evaluated as

$$W = \sum_{\text{cell } k} \left\{ w_k \lambda_k (1 - B_k) + c_k B_{hk} \sum_{\text{cell } j} v_{jk} \right\}, \tag{5.4}$$

where v_{jk} is calculated with Equation (5.2), λ_k is the new calls offered to cell k, w_k is the revenue generated by accepting a new call in cell k, c_k is the cost of dropping a handoff call in cell k, and B_k and B_{hk} are the new call and handoff blocking probabilities, respectively. In Figure 5.14, we can see the effects of

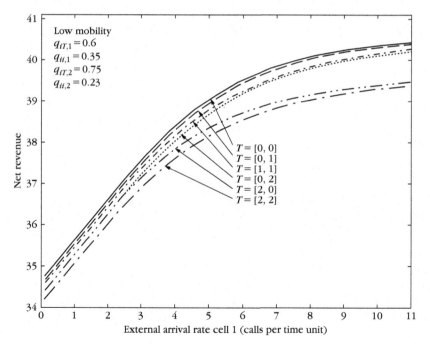

FIGURE 5.14

Network net revenue for low mobility and reservation.

mobility and reservation in net revenue. The reservation helps give priority to handoff calls, and as this increases, net revenue decreases.

The model has also been considered in evaluating vertical handoffs in a wireless ATM network in Garca-Berumen and Vargas [25], where mobility also is an important aspect of capacity. An extension of the model to consider out-age effects together with mobility is presented in Gallegos and Vargas [23]. Among the advantages of the model using the Jackson network approach, we find the low computational complexity to be implemented, the easy way to modify mobility by varying the routing probabilities, and the mobility effects captured through the traffic equations in Equation (5.2). Disadvantages include the lack of clear sensibility to holding times, which turn out to be a fundamental part of the definition of offered traffic in erlangs, and the difference between cell residence times and call holding times, which are dependent upon the routing probabilities. Examples of works that discuss at these times in cellular networks with mobility are Orlik and Rappaport [61] and Vargas-Rosales et al. [79]. Both use a multidimensional Markov chain. The main difference is that in [79] call holding times are given by the sum of several exponential random variables for calls undergoing handoffs so that in general erlang-k distribution is obtained. In contrast, in [61] general distributions are used.

Vertical handoffs among several wireless providers have also been treated as an application of the simple model. See Mora-Zamorano and Vargas [52] for a model where resources are shared via random and sequential strategies, and Vargas-Rosales and Stevents [80] for the case in which resources are shared in an adaptive way by considering a state-dependent Markov chain, which helps to obtain better results than those in Miranda-Guardiola and Vargas-Rosales [52] due to its adaptability to traffic overload. In this case, the least-loaded resource is the one chosen to accommodate users offered as handoffs to other networks or technologies. The use of the least-loaded algorithm guarantees that idle infras-tructure will be used and will generate revenue. In the aforementioned papers, a rate of return function is formulated that considers the carried traffic in the network, and a nonlinear optimization problem is solved where the maximum offered rates are obtained in terms of several degrees of mobility. Reservation is also used for handoff calls.

The model can be applied to a multicarrier system with the same mobility and offered traffic conditions. This could be the initial step toward modeling of the vertical handoffs that need to be performed in a 4G network. The main goal of the multicarrier system is understood as the organized integration of carriers to ben-efit from all idle resources. The use of reservation needs to be carefully evaluated to obtain the trade-off required. In Figure 5.15, we can see the network net rev-enue when vertical handoffs are allowed even for incoming new calls that find no free channel. We can see that the use of reservation, in both cases of mobility,

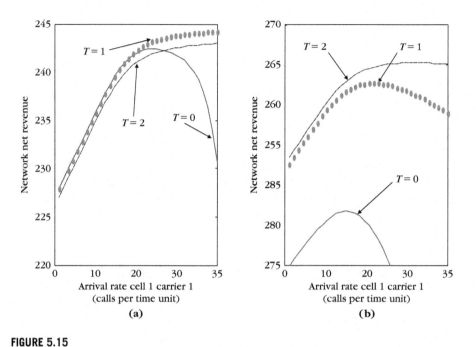

FIGURE 5.15

Network net revenue for (a) low mobility and (b) reservation.

benefits net revenue since it keeps increasing as offered traffic increases. In contrast, we can see how net revenue decreases as offered traffic increases when no reservation is used. We can also see that reservation needs to vary according to mobility levels; that is, it needs to increase as mobility increases.

When we want to change the scenario toward the reconfigurable networks, we need to consider different issues regarding mobility. Because we lack an infrastructure, the use of location areas and cells with paging and control messages has no application. A general scenario will need mobile nodes to act as beacons or references to the remaining nodes in the network. These reference nodes can be fixed or mobile. The issues of modeling mobility are important since they play important roles in performance when introduced in more general models to predict or evaluate link establishment and maintenance that affect topology management. Some of these models can be simple, such as a square area where nodes move like billiard balls in the area, and when an edge is reached, two options are used—one where a complementary angle defines the new direction of movement and another where the node *appears* suddenly on the other side of the square region. In both cases, the nodes might be moving at constant speed with a direction determined randomly in $[0, 2\pi]$. Also, it can be more complicated by considering a speed varying randomly as well within an interval $[0, v_{MAX}]$. Another way would be to have a discrete simulation where at each discrete event, nodes randomly select velocity, direction, and time in discrete slots through which they will travel at those values.

Simulations are important since it is easier to achieve better modeling of node movements with more realistic behavior as discrete events are executed. In general, one of the mobility models widely used is the random waypoint [44], which includes as a special case the one described as the billiard model in the previous paragraph. The existence of a mean trip duration implies that the distribution of node mobility converges to a stationary distribution (see Le Boudec and Vojnovic [44] and Navidi and Camp [58]). This model is simple and tractable for implementation in simulations together with other aspects of networking. One of the disadvantages is that it does not capture realistic behavior of nodes, especially in contrasting scenarios such as indoor and outdoor areas, since mobility will have different degrees of freedom as scales change [45].

A generic framework that intends to present a mobility model that integrates heterogeneous wireless networks with vertical and horizontal handoffs is introduced in Zahran and Liang [83]. The introduction of distributions of the phase type is also an important aspect of this framework. The model tries to integrate technologies such as 3G, WiMAX, and WiFi by carrying out vertical handoffs that are related to cell residence times for cellular technologies and zone residence times for other technologies. The use of phase (PH)-type distributions for residence times fitted to mobility traces is also discussed. The advantage in terms of analysis is that PH-type distributions have simple results when superpositioned, giving another PH-type distribution. Examples of PH-type distributions are the hypoexponential and hyperexponential distributions, and the Coxian random variable, which gives rise to the Coxian model [14].

Simulations for a network integrating 3G and WLAN technologies are presented in [83]. Evaluation is in terms of cellular session utilization and the rate of vertical handoffs. These measures are also evaluated analytically to find good agreement with the simulations. The models used include the one based on the Coxian model [14] and a zone residence time determined by PH distributions. The mobility models just discussed where zone and cell residence times are considered can be classified as macroscopic mobility models since they describe large-scale user mobility. When one wants a better representation of mobility at small scales such as that encountered in reconfigurable networks, microscopic models are the best choice [45].

In [45], mobility modeling is carried out via an approach known as behavioral mobility (BM), where modeling concentrates on the realistic representation of mobility as a function of behavioral patterns. A discussion of models, such as the Markovian, which forms part of the macroscopic classification, is introduced, where applications of cellular networks are straightforward and the level of detail is usually low. Also discussed are models comprising a mixture of micro and macro that have applications to the representation of mobility in reconfigurable networks with medium granularity. These models include the random waypoint mobility model. An enhancement of that model is also discussed as an example of a microscopic mobility model with high accuracy and applications in reconfigurable networks. Pedestrian-level and group-mobility models are also introduced in Legendre et al. [45].

5.2.3 Dealing with Mobility

The mobility concept is basically the same one used in cellular networks where the system is organized in groups or clusters of cells to form location areas. The system periodically will order the mobiles to update their information, thus producing data that help to determine the location of such nodes within a certain accuracy range since a location area and a cell and sometimes a sector are known for each user. If one wants to provide a user's absolute or relative coordinates, any of the techniques discussed in Chapters 2, 3, and 4 are useful. Some concepts can be based on deployment area organization with the use of attractors to indicate locales where user density increases at certain periods of time during the day (e.g., stadiums, malls, and college campuses).

We are interested in predicting the position and mobility of cellular customers, and in determining other parameters like blocking probability in a cellular system. We use the concept of social grouping behavior as an open-and closed-migration process with transition rates given by Kelly [40]:

$$q(\mathbf{n}, T_{.k}\mathbf{n}) = \Psi_k(n_k),$$

$$q(\mathbf{n}, T_{j.}\mathbf{n}) = \phi_j(n_j), \qquad (5.5)$$

$$q(\mathbf{n}, T_{jk}\mathbf{n}) = \lambda_{jk}\phi_j(n_j)\,\Psi_k(n_k),$$

where $\phi_j(n_j) = d_j n_j$, and $\Psi_k(n_k) = a_k + c_k n_k$; a_k is the attractiveness to an outside; user of belonging to group k; c_k is the attractiveness to an outsider of being an individual in group k; d_j is the propensity of an individual to depart from group j of an individual in group j; λ_{jk} is a measure of the mobility of a user from groups j and k; and n_k is the number of active users in attractor k.

Attractor

We consider a city to be a finite associated group of zones of activity called attractors, where the number of customers varies according to an attractor's characteristics and its relationship with other adjacent attractors. An attractor is a site that attracts customer movement and at which they remain for a given time (e.g., work areas, residential areas, entertainment areas, shopping centers, theaters). We define the attractor according to cellular customers; in order to determine the attractors present in a certain zone or cell, we have to analyze the customers, hour of the day and geographical area, and characteristics and interrelation with other attractors. The characteristics of each attractor used in Equation (5.5) now become dependent on time; that is, $a_k(t)$, $c_k(t)$, $d_k(t)$, $\lambda_{jk}(t)$, and $n_k(t)$.

We consider a set of J attractors, but we shall allow customers to enter and leave the system as well as to move between attractors; thus $T_{j\cdot}$ represents a customer leaving the system from a attractor j, $T_{\cdot k}$ represents a customer entering the system to attractor k, and T_{jk} represents a customer moving from the attractor j to k. We also introduce another factor into the equation that we define as geographical feasibility, which is the parameter that indicates how easily customers can move from one attractor to another as shown in Figure 5.16. The parameter $\lambda_{jk}(t)$ allows us to measure the mobility of customers from attractor j to k, and is given by

$$\lambda_{jk}(t) = \frac{n_k(t)\kappa_{jk}}{n_j(t) + \sum_{i,i \neq j} \kappa_{ji} n_i(t)}. \tag{5.6}$$

Then we can predict the number of active customers in attractor j at time t, $n_j(t)$, by

$$n_j(t) = n_j(t-1)\left[1 - \sum_{i \neq j}\lambda_{ji}(t)\right] + \sum_{i \neq j}\lambda_{ij}(t)n_i(t-1). \tag{5.7}$$

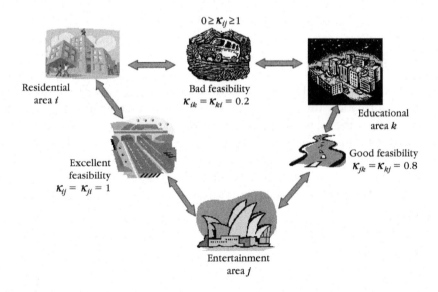

FIGURE 5.16

Feasibility between attractors.

This simple model helps us organize a network into areas where feasibility parameters and mobility are considered in order to predict user numbers. The number of users can be employed to calculate the amount of traffic expected to be offered to the network in erlangs, which in turn will determine the number of channels needed at a certain grade-of-service level. Base stations must be close to attractors with greater population than others. Further results with this mobility model are in Baca [85] and Bermudez [86].

5.2.4 Mobility and Location-Based Services

In current wireless networks such as those in 2.5G and 3G, location-based services (LBS) are an important part of new application and service development. From the simple search for a restaurant or a store within an area to emergency services, LBS are an essential part of these networks and form a fundamental block to take the technology to the next level.

LBS can be organized according to location areas, cells, and sectors within a cell so that applications such as the reception of short messages announcing sale events in stores and instructions to follow a map in order to get to a destination can be implemented. Accurate location of the user receiving the service is in general not needed or, as we say, service dependent. Mobility in these services plays an important role, since the handoffs must handle extra variables to continue providing the LBS if such is active at the time the handoff takes place. Information regarding the application can be organized to look ahead for possible problems regarding mobility. For example, when the service is concerned with finding directions to a specific destination, the system can prepare itself to receive handoffs as the instructions are delivered, reserving resources where needed.

In other areas, control packets can be sent to users in order to update their location information through extra base station elements that are identified with certain locations; when the user carries out the update, the system will know that it is close to certain commercial areas and messages regarding services or local information can be delivered. City traffic monitoring is another important application that could be delivered through the use of LBS. In cities with video cameras recording street traffic, users can request this information, especially with 3G where low-quality video exists. In airports, packets could be delivered to users in the vicinity to inform them of flight departures and arrivals.

One should not confuse LBS with services that ask users to send a message in order to receive sports news or horoscope predictions. These are not LBS, although some could become LBS—for example, students at a university campus could ask for their grades through an LBS, or announcements could be made of courses to be offered during the semester or extracurricular activities, or faculty could use podcastings to deliver information to students.

Mobility is an important issue from any network point of view, including the operator, service provider, manufacturer, and end user. For some, LBS mobility plays an important role because resources in the network have to be managed in order to continue the philosophy of avoiding handoff drops. If reservation must be used, it has to be defined within algorithms that see local information but use it at a network level to optimize the use of resources. Today, we consider new paradigms, such as cognitive radio, where adaptation is a key feature of the devices through the use of a software-defined radio (SDR). This paradigm enables the prompt adaptation of transceivers that allow the application to use LBS. The system must also obtain user location information to provide access to information in a personal way and to support location awareness as well as positioning of users to achieve true LBS at all granularity levels.

5.2.5 Mobility and Location

Most of the treatments and algorithms for PL are focused on scenarios where networks are static—that is, nodes are not mobile. Some proposals have suggested the use of the algorithm repeatedly over a discrete-time definition of the events in the network, allowing for updates of information that are useful for the location process, such as neighboring nodes, routes, hop distance, and distance toward landmarks or reference nodes. Even if the algorithm were repeated over a discrete-time evolution, in a heterogeneous network, one could not ensure that information was updated properly in order to perform positioning, since processing delays and transmission power could have important roles in this scenario.

When nodes are mobile, updates are triggered often due to topology changes and received signal strength decay. Thus, even if a node is able to obtain the processing power needed to carry out updates, the parameters involved in such updates might be compromised due to wireless channel and propagation impairments. Mobility increases this effect by causing Doppler effect and delay spread, together with the changing of the multipath behavior of the channel. At high speeds, coherence time and coherence bandwidth are affected causing information to travel over different channel states with uncorrelated effects that make message recovery difficult.

When a cellular scenario is considered, we have the effects just mentioned, which should be captured by the location algorithm through the statistical estimation of channel parameters used in the algorithm. Another helpful procedure is to perform the signal strength signature as shown in Section 4.1.2. The effects due to mobility of the node of interest in a cellular scenario also affect algorithms based on traditional methods such as AOA, TOA, and TDOA. For example, when a signal arrives at the point of measurement, DOA or TOA can be determined somehow, but once the parameter has been determined, the node could have already moved. Thus, such parameters will have noise not only due to the environment but also due to mobility and inaccuracy will increase. In a scenario such as the cellular, mobility will not necessarily be accompanied by an increase in the number of reference sites or landmarks that are able to provide support in the positioning algorithm.

In contrast, in a reconfigurable network scenario such as in ad hoc and sensor networks, mobility of the node of interest or the surrounding nodes could help increase the number of reference nodes used to improve accuracy of the positioning algorithm. The trade-off would be in the time needed for the network to be ready to perform such algorithm. In other words, a topology update needs to be carried out, and when the information update has been obtained, the location algorithm could be performed. Another bottleneck is the processing time of the nodes depending on their hardware characteristics. When the node of interest is an active node, such as a sensor, and it is moving, the problem of locating and tracking in real time needs to be addressed. A possible solution could be the use of collaborative algorithms that help improve performance. The node could be moving around an area with nodes that could help by serving as landmarks that measure parameters from the moving target. These parameters could be reported and averaged over the area traversed.

Since mobility in reconfigurable networks has been shown to help improve capacity in terms of throughput [29], it is also possible that it could help improve location accuracy if performed with a collaborative method since landmarks could be cooperating by sharing information that helps to locate not only the target, but also other nodes in the network [24]. Reconfigurable networks can have several integrated scenarios, including fixed reference and target nodes, mobile references, or mobile targets, or both. In any case, since topology changes, signaling is increased even in those portions of the network where no changes have occurred, because propagation by flooding of the new topology information occurs. The idea could be to have topologies organized in clusters so that moving nodes could update their information with cluster partners and obtain information and localization from those partners. If information is needed outside the cluster, then it is shared by intercluster routing as requested. Reference nodes that move and are landmarks could have a positive impact by closing in on a node of interest by improving accuracy of the measurements involved in the positioning algorithm [60].

5.3 TOWARD THE CR PARADIGM FOR POSITION LOCATION

Cognitive radio (CR) is a new paradigm that emerged from studying the problem of spectrum scarcity or the efficient use of the RF spectrum. It was found that large portions of the spectrum are unused for much of the time. The main idea is for intelligent devices to sense the spectrum to find and use unoccupied portions without causing interference affecting other authorized users. The device would also be able to perform frequency hopping in order to leave the frequency free to primary users assigned to such an RF band, or it could modify its transmission characteristics (power, modulation, coding, etc.) to coexist with such primary users. These ideas were proposed for the first time in Mitola [50]. In this section, we present the concept of CR together with some recent results in the area of PL. We also present selected technologies that will certainly ease the process of achieving this new paradigm.

5.3.1 The Concept of Cognitive Radio

Even though a clear definition of CR has not been recognized, we can point to certain concepts produced in recent years and separate from a cognitive radio feature.

For example, CR as a paradigm for wireless communications represents the use of devices in a network where both the devices and the network could change or adapt their characteristics to achieve higher efficiency levels. The particular adaptation takes place based on sensing the environment variables of interest such as RF spectrum occupancy, traffic, and location awareness.

We can see that the devices in this paradigm will *learn* to adapt according to their sensing of the environment in order to maintain or improve quality of service (QoS) and reduce interference to other devices or services. To achieve such paradigm and device features, two avenues have been used (i.e., SDR and spectrum sharing).

The idea behind spectrum sharing is to organize spectrum access in order to assign resources on demand and location bases. Also, the process could be managed by the network and end users instead of regulatory entities. The Federal Communications Commission (FCC) reports that in the United States a large percentage of the spectrum that has already been allocated is underutilized. Sharing is seen as a possibility with the new paradigm due to device adaptability to several of the environment parameters such as bandwidth, frequency, power, interference, coding, and synchronization. The device will also be able to use this under utilized spectrum if it also adapts itself in terms of technologies, packet formats, packet lengths, and protocols. This self-adaptability derives from the concept of SDR, as device characteristics are changed from the *inside* [51].

The success of spectrum sharing will come together with methods to achieve almost no interference to primary users in the selected frequency band. The quantification of such interference is also an issue, and a performance measure to achieve this has been proposed with the definition of *interference temperature*, in Haykin [33], where issues such as cooperation, power control, and dynamic spectrum management are also discussed. CR needs certain capabilities to reach such goals, among which location plays an important role of sensing the specific environment. At some spatial points, for example, a CR is aware of technologies such as cellular coverage, range of WLAN signals or areas defined by Bluetooth technology. Hence, a CR has sensors for location information with learning and intelligent algorithms to adapt itself by reconfiguring according to the environment. The disadvantages that are evident at this point in time are the complexity of the hardware necessary to achieve the sensing, complexity and computational requirements to achieve intelligence, and communications cost incurred in the sensing, learning, decision-making, and reconfiguring phases. Research is currently focused on CR combined with multiuser detection, ultra wideband (UWB) communications, and cooperative diversity. A survey of these mixes can be found in Glisic [26].

As discussed previously, awareness is important for CR, especially location for the device to use local parameters to sense the environment and reconfigure itself to adapt to changing conditions. Realization of these tasks is known as location awareness [8, 9]. Application of these location-awareness techniques is in LBS, characterization of the environment, network optimization assisted locally, and transceiver optimization. In addition, a location awareness architecture is introduced in Celebi and Arslan [8], where the estimation of position is a relevant issue in order to carry out the CR paradigm.

Positioning applications in CR systems can help in monitoring the environment to conduct dynamic spectrum access, which is fundamental to achieving optimization at the network level. Also, traffic monitoring can help provide LBS services to traffic commuters, and the sensing of some parameters will make the device adapt to channel conditions by changing its transceiver. It is important to note that CRs will form cognitive wireless networks that will work cooperatively with other technologies such as Bluetooth, CDMA, WiMAX, and satellite. The issue is to adapt in order to achieve interoperability. The potential of location techniques for CR devices is higher since the latter can use traditional techniques, such as those discussed in Chapters 2, 3 and 4, based on the estimation of position, or new sensing techniques and scene analysis; some postprocessing such as video and image information can also be used. A cognitive radio positioning system with two modes is also discussed in Celebi and Arslan [8], where bandwidth determination and dynamic spectrum management are used. Considering a channel with an estimated distance error of σ, an observation period containing K symbols, a signal-to-noise ratio of γ, and a signal propagation speed of v, we can determine the required effective bandwidth β as follows [8]:

$$\beta = \frac{v}{\sqrt{K}\gamma\sigma}. \tag{5.8}$$

This required effective bandwidth is used by the dynamic spectrum access to find and provide to the CR system a bandwidth to carry out the positioning algorithm; an estimate of the location is obtained afterward. There are also applications of location awareness such as location-assisted handoffs and location-assisted network optimization, discussed as well in Celebi and Arslan [8].

5.3.2 Multiple Antenna Systems

Multiple antenna systems have been extensively used due to the advantages of diversity, especially of the spatial kind. Currently, demand for high-data-rate wireless communications is still growing rapidly. In the past technologies with the greatest media hype promised more value to the end user, but so far success has not been attained. Limiting factors in wireless environments are the lack of wide network coverage, vehicular mobility, and inadequate spectral resources. For many years now, multiple antenna techniques—especially multiple-input, multiple-output (MIMO) technology—have gained much research attention. MIMO is very promising for technologies such as 4G and 802.11n wireless LANs, and it will probably be deployed in environments where several users require high downlink data rates at single base stations. We provide an introduction to the topic by developing discussion from the traditional single-input single-output case to the MIMO case. In this section, we provide a brief summary of MIMO systems.

For wireless services demand will likely continue to grow due to multimedia services for mobile terminals such as live video streaming, videoconferencing, mobile games, and mobile television. These services not only require high data rates from wireless networks but also support implementation of QoS schemes.

Development of higher and higher data rates has been gradual. The 2.5G technologies included high-speed circuit-switched data (HSCSD) and general packet radio service (GPRS) [6], of which the former is able to provide 57.6 Kbps in good radio conditions using four time slots, and the latter is, depending on the channel coding, capable of rates up to 114 Kbps. However, the GPRS has been offering peak data rates of only 53.6 Kbps (4 times 13.4 Kbps) in commercial networks, and overall performance is even lower due to packet retransmissions and connection establishment (i.e., reserving time slots from the network).

After introducing 2.5G technology and while standardization work for 3G systems such as the Universal Mobile Telecommunication System (UMTS) was going on, a bolt-on enhancement for GSM systems called enhanced data rates for GSM evolution (EDGE) was introduced by 3GPP—also responsible for GSM and GPRS. The primary target group for EDGE is the operators, who lost the competition of large frequency bands for 3G technology. EDGE is able to provide 236.8 Kbps with four time slots in perfect radio conditions.

As deployment began, the 3G technologies became available in an increasing number of countries. For example, in Mexico the CDMA-2000 standard has been deployed providing services such as TV. During the auction of frequency bands in Europe, there was enormous media hype about the 3G/UMTS technology with promises of videoconferencing and other applications. However, as the implementations lack QoS features and the pricing of data transfer is still high, the popularity of 3G/UMTS is growing at a much slower pace than expected, and there is still a great void to be filled with mobile services using the faster transfer rates of 3G (up to 384 Kbps in commercial WANs). Additionally, maximum goodput is achievable only with contents, which require longer download times and somewhat perfect signaling conditions. To increase data rates even more, 3GPP has defined a technology called high-speed downlink packet access (HSDPA) in its UMTS/WCDMA standard. It promises data rates exceeding 10 Mbps [57].

With this short historical review, it is clear that there is still a great need for faster transmission speeds in WANs and WLANs. Despite huge effort, videoconferencing and mobile TV still require faster Internet access for the end user. MIMO technology is a very promising answer to this demand. It is often suggested as the basis not only of future 4G technology in WANs, but also of the new WLAN technology with over 100-Mbps Internet access—referred to as the 802.11n standard. The rest of this section discusses the wireless channel in general with the single-input single-output (SISO) case, and step by step we develop the discussion toward the MIMO case. The complete summary is in Perala and Vargas [87].

Wireless SISO Channel

We begin with the basic concepts of the common wireless channel, which differs from the wired channel because of such phenomena as multipath and interference. Let us first consider the simplest case, where we have a time-invariant channel. When we send signal $s(t)$ over a channel with impulse response $h(t)$, the received signal has the following expression:

$$r(t) = h(t) * s(t) + n(t) + j(t), \tag{5.9}$$

where the $*$ operator means convolution operation, $n(t)$ is additive white Gaussian noise (AWGN), and $j(t)$ is the interference from other users. The channel usually varies over time in stochastic fashion and attenuates the amplitude of our signal; thus, we define it as

$$h(t) = \alpha\delta(t-\tau), \quad \alpha \in [0, 1], \tag{5.10}$$

where α is the attenuation coefficient of the channel, and $\delta(t-\tau)$ is the Dirac's delta function with path delay τ. Now we can combine these equations to obtain

$$r(t) = h(t) * s(t) + n(t) = \int_{-\infty}^{\infty} h(t)s(t-\tau)d\tau + n(t), \tag{5.11}$$

and in the frequency domain we have $R(f) = H(f)S(f) + N(f)$. Note that here we do not comment on the nature of the signal $s(t)$, and therefore it can be chosen by the designer of the transmitter-receiver. In addition, the interference $j(t)$ has been ignored thus far.

The multiuser environment makes the transmission dilemma even more difficult because it adds interference to the system. For certain users, the signals of other users are seen as undesirable. In the mathematical sense, although very generally speaking, the multiuser environment in the frequency domain for K users is described as

$$R(f) = \sum_{i=1}^{K} H \cdot S_i(f) + N_i(f), \tag{5.12}$$

where we can separate the desired user as

$$R(f) = H \cdot S_1(f) + \sum_{i=2}^{K} H \cdot S_i(f) + N_i(f) \tag{5.13}$$

$$= H \cdot S_1(f) + J(f) + N_i(f),$$

where the term $\sum_{i=2}^{K} H \cdot S_i(f)$ is the aforementioned interference $J(f)$. This multiuser aspect is valid also for multiantenna cases. Considering multipath propagation, the channel impulse response is given by

$$h(t, \tau) = \sum_{k=0}^{n} \alpha_k(t)\delta(t-\tau_k)e^{-j\theta_k}, \tag{5.14}$$

where k denotes the index of each multipath component and j is the imaginary unit. If these components with various phase delays are summed in the receiver, they cause multipath fading.

The worst-case scenario would be to have two components with 180-degree phase difference summed, in which case the sum is seen by the receiver as zero.

This phenomenon can be overcome somewhat with several techniques—for example, with the RAKE receiver used in CDMA systems [67, 68]. Mobility of the transmitter and/or receiver increases the complexity of the channel model because we need to take into consideration the Doppler shift. Often it is assumed that the user is stationary or moving very slowly, which is reasonable because typically the users of WLAN technology use their laptop when sitting somewhere and movement usually takes place at walking speeds.

When all the information discussed in this section is combined into a single-channel model taking into consideration attenuation, path delay, noise, phase shift, fading, multiuser environment, and discrete-scatter Doppler phenomena, we get our channel model,

$$h_i(t) = \sum_{k=0}^{n} A_k \delta(t - \tau_k) \cdot e^{j2\pi v_k t} + n_k(t) + j_k(t), \tag{5.15}$$

where A_k is the term that represents all those attenuation and phase-shift effects on the signal in each of the n multipath components, and v_k is the Doppler shift [68].

SISO is the traditional way of seeing multipath propagation. In SISO systems, we have only one transmit antenna and one receive antenna, and therefore we are not able to send multiple data streams to the channel and/or receive multiple copies of the signal on the various antennas, as in the multiantenna systems discussed in the following sections.

Wireless SIMO Channel

The difference between SISO and the more sophisticated, but still classic, approach called single-input multiple-output (SIMO) is, of course, multiple antennas on the reception (Rx) side. With more than one antenna at the Rx end the system gains benefits in spatial diversity. This is achieved by carefully combining the signal copies in the receiver. The price we pay for improved bit error rate (BER) and signal-to-noise ratio (SNR) is more complexity in the receiver [27, 68].

In the SIMO case, we send only one signal $s(t)$ to the channel, but we receive the same signal (and multipath copies) at multiple antennas. As an illustration, consider a system with n receive antennas. The channel for each one of the n subsystems is given by

$$h_i(t) = \sum_{k=0}^{n} \alpha_k(t) \delta(t - \tau_k) e^{-j\theta_k} \overset{F}{\longleftrightarrow} H_i(f), \tag{5.16}$$

which gives the following for the vector of received signals:

$$
\mathbf{R_n} = \begin{bmatrix} R_1(f) \\ R_2(f) \\ \ldots \\ R_n(f) \end{bmatrix} = \begin{bmatrix} H_1(f) \\ H_2(f) \\ \ldots \\ H_n(f) \end{bmatrix} \cdot S(f) = \mathbf{H_n} \cdot S(f). \tag{5.17}
$$

Signal power fluctuates in wireless channels randomly; in other words it fades. By using several receive antennas, we achieve diversity, which is a good technique to combat fading. Diversity benefits derive from transmitting the signal over multiple (ideally) independently fading channels. It is possible to separate channels in time, frequency, or space. Spatial diversity is preferred, since it doesn't require more bandwidth or time. If the transmitted signal is suitably constructed, the receiver is able to combine the received signals in such a way that amplitude variability is significantly reduced. The diversity techniques do not directly increase the data rates, but they improve the quality of the received signal, and therefore they make it possible to use other techniques (e.g., modulation schemes) to increase data rates [64].

Wireless MISO Channel

Multiple-input single-output (MISO) is a more recent approach in spatial diversity techniques than the aforementioned SIMO. The situation is somewhat the opposite of the SIMO case, since now there are multiple antennas at the transmitter side and only one at the receiver side. Similarly, as in the previous section, based on Equation (5.16) we can write the channel matrix for the MISO channel. The channel matrix $\mathbf{H}$ is similar as before, but the difference is comprised of the signal vector $S(f)$, since we are able to break up the high-speed data stream into m lower-speed streams and transmit them over the channel. Thus, we obtain

$$
\mathbf{R_n} = \begin{bmatrix} R_1(f) \\ R_2(f) \\ \ldots \\ R_m(f) \end{bmatrix} = \begin{bmatrix} H_1(f) \\ H_2(f) \\ \ldots \\ H_m(f) \end{bmatrix} \cdot \begin{bmatrix} S_1(f) \\ S_2(f) \\ \ldots \\ S_m(f) \end{bmatrix}^T = \mathbf{H_m} \cdot \mathbf{S_m^T}. \tag{5.18}
$$

In addition to diversity gains and spatial multiplexing gains, MISO systems introduce a new benefit, namely coding gain. It can be achieved in the context of transmit diversity by appropriately designing the transmitted signals, resulting in space–time codes (STCs). Coding gain is achieved with no bandwidth efficiency loss [64].

Wireless MIMO Channel

MIMO systems combine MISO and SIMO systems as one and therefore reap the benefits from both. We have two antenna arrays, one at the transmitter and one at the receiver, respectively. Following the same procedure as before, we can obtain the more general expression for m transmit and n receive antennas as follows:

$$\mathbf{R_n} = \begin{bmatrix} H_{11} & H_{12} & \cdots & H_{1m} \\ H_{21} & H_{22} & \cdot & \cdots \\ \cdots & \cdot & \cdot & \cdots \\ H_{n1} & H_{n2} & & H_{nm} \end{bmatrix} \cdot \begin{bmatrix} S_1 \\ S_2 \\ \cdots \\ S_m \end{bmatrix} = \mathbf{H_{nm}} \cdot \mathbf{S_m}. \tag{5.19}$$

MIMO systems offer several gains compared to single-antenna systems. Diversity, spatial multiplexing, and coding gain have already been discussed briefly in the SIMO and MISO sections, and the additional gains are array gain and interference reduction. Array gain is realized through processing at the transmission (Tx) and Rx ends and it results in an increase in average receive SNR due to coherent combining effect. Co-channel interference is reduced by using the differentiation between the spatial signatures of the desired channel and co-channels, but knowledge of the desired channel must be available [64].

Capacity of MIMO channels has been studied for several years now and roughly two categories of design may be used [2, 57]:

- When full channel state information (CSI) is available (i.e., amplitude and phase response for each channel) on the Tx side, eigenvector steering (ES) may be used to approach full capacity of the MIMO channel.

- When partial CSI is available on the Tx side, receiver processing is used to separate the various spatial streams and spatial spreading.

The capacity of ES with full CSI has been discussed in detail by Andersen [2], and we just note that the spectral efficiency for the $M \times N$ MIMO system is given by

$$C = N \log_2 \left(1 + \frac{P}{N}M\right), \tag{5.20}$$

where M is the number of transmit antennas, N is the number of receive antennas, and P is SNR for one Gaussian channel. As discussed before, MIMO channels may have very high spectral efficiency in environments where rich scattering exists. The scattering enables the arriving signals from each transmitter to be highly uncorrelated [32]. However, if the arriving signals are correlated, high spectral efficiency is reduced. The separation of antennas reduces the correlation, but in certain circumstances the performance is deteriorated significantly even if the signals have zero correlation. This effect is referred to as a "keyhole" or a "pinhole," and has been studied both theoretically and empirically in Chizhik et al. [11]. The keyhole appears, for example, in such physical environments as tunnels and hallways, and it is nothing more than severe multipath fading caused by specific environments.

In this section, we discussed the attributes of wireless channels from the multiantenna systems point of view. We provided simple channel models for the SISO case and applied it in more complex antenna array scenarios. Multipath propagation is a key concept in understanding fading and the many relative benefits of MIMO systems. Such concepts comprise an active area of research, and the basics of user movement and multiple users were discussed. We also provided a short introduction to capacity of the wireless MIMO channel and the performance-deteriorating keyhole effect.

5.3.3 Basics of Cross-Layering for Reconfigurable Networks

Traditionally, networks were designed with a layering architecture in mind, where functions were separated and organized to offer services to upper layers and use services from lower layers. Open System Interconnection (OSI) is the standard that shows the hierarchy followed for this architecture. In contrast, reconfigurable networks impose new challenges to network designers, administrators, and operators, since the definition of functions and operations in a layered architecture is not as clear as for wireline networks. For example, it is well known that the TCP protocol offers a connection-oriented service and works on an end-to-end basis, reacting to congestion and making the transmitter reduce or increase the packet rate according to measures of round-trip delay. In general, the effects on round-trip delay were due to high levels of congestion through the paths that the information traveled, but in networks with wireless links, this might not be true since delay might be caused by packet buffering because of conditions such as outage or interference. With this, TCP would calculate that congestion was taking place when a very different phenomenon was present. Another example would be the application of routing algorithms, since these depend on the construction of a topology that depends on the status of each individual link; thus, physical layer phenomena affect algorithms of other layers, which did not occur in wireline networks. For this reason, it is important to integrate these functions and operations so that they can interoperate in order to provide more efficient services. This is what we call *cross-layering*.

In general, the treatment of these effects that traverse several layers can be formulated through the use of objective functions that need to be optimized, using restrictions where each of the effects to be studied is introduced as a constraint. A good discussion of some of these issues, as well as the formulation of challenges and new paradigms within cross-layering, can be found in

Van Der Schaar and Shankar [78], where adaptation techniques and optimization methods are introduced to take advantage of cross-layering. Issues concerning fairness and competition are also briefly discussed as new paradigms that should be considered in the cross-layering of reconfigurable networks.

Cross-layering topics appearing in the literature include, for example, the idea of using functions in the data link layer such as the automatic repeat request (ARQ) to improve physical-layer performance and vice versa [18]; these improvements are obtained by using new multiple access mechanisms in the data link layer that provide network diversity. They also show that the application of cross-layering allows throughput of almost one packet per slot and small delays over several traffic-load ranges. The disadvantage is the increased complexity of the receiver and greater power usage. In Berry and Yeh [4], resource allocation is considered within the context of cross-layering; resources include the transmission power and rate of each user. Measures of QoS at the network level and physical-layer performance are obtained. They use a flat-fading Gaussian multiaccess channel with bandwidth W, noise density $N_0/2$, fading level h, and rate r to obtain the minimum power required to transmit at rate r that is less than the channel capacity, as follows:

$$P = \frac{N_0 W}{h} \left(2^{r/W} - 1 \right). \tag{5.21}$$

This approach is extended to the multiaccess case to solve for the powers of each user with the rates lying at the extreme points of the capacity region. In De et al. [15], cross-layering is used in the context of reconfigurable networks for the case of sensors using CDMA for multiple access. Grids and hexagonal topologies are compared using a uniformly distributed user population. The cross-layering is applied by studying the physical layer constraints, such as BER, on performance measures at the data link and network level, such as throughput.

5.3.4 Cooperative and Collaborative Wireless Networks

The use of multiple antennas such as the examples in the scenarios SIMO, MISO, and MIMO is a key feature in the concept of cooperation. When cooperation needs to be implemented in a reconfigurable network, such diversity can be obtained by coordinating several transmitters or receivers in order to achieve the same advantages as those for the MIMO channel (i.e., diversity). Another important issue to consider when looking into cooperative schemes is that not only physical layer performance must be measured, but also the effects and advantages of the coordination at higher layers to improve performance in an integral way in the network architecture.

In order to implement cooperative communications in reconfigurable networks, nodes must be organized in some logical way so that they recognize potential helpers as a cluster. Such a group of nodes can be commanded by an AP that can be fixed or mobile but has more processing capability; or the group of nodes could be coordinated by the node in need for a certain period of time during which cooperation must take place. In Scaglione et al. [72], a discussion of these fundamental principles is introduced, where BER and outage are chosen as performance metrics. They compare the noncooperative transmission versus the cooperative scenario, considering in principle a synchronous system with the amplify-and-forward strategy, as well as a comparison with the decode-and-forward strategy. It is shown that cooperation has advantages over noncooperative methods. Also, they discuss the issue of cooperation as a networking strategy to improve performance with influence and effects on lower layers, such that the concept of an end-to-end link remains to be well defined.

In Hong et al. [35], cooperative schemes with the objective of improving performance by relaying information through several nodes in the network and optimally allocating power and bandwidth are introduced. In order to achieve such cooperation, channel state information is used. Hong et al. also recommend defining strategies to consider the channel state information estimation improvement to achieve better power allocation. Cooperation in multihop wireless networks is still an open issue to be studied and needs to be considered within the cross-layering framework together with the end-to-end link abstraction. Cooperative schemes increase the communication cost in the network, and strategies to decrease this cost as well as their robustness to time-varying characteristics, such as network topology, need to be investigated.

In general, nodes in the network must be able to decide by themselves whether to accept a relaying request. This would lead to a noncooperative or cooperative scheme depending on the decision taken by those nodes. The decision-making process must depend on variables such as environment and internal energy, since ultimately, it will determine the degree to which relaying can take place. In Srinivasan et al. [73], a network with this idea of considering remaining energy in the nodes is introduced, and an algorithm to determine the optimal proportion of cooperation that each node should receive is formulated. The results show that such an algorithm converges to the optimal operation point. The remaining energy is considered by dividing the nodes in the network into classes where each class has an energy constraint and an expected lifetime. The results are presented by formulating a game theory framework for the algorithm.

The notion of topology organization is also helpful in establishing cooperative algorithms in reconfigurable networks. Nodes can be organized in clusters, and then each of those groups can determine which nodes will serve as intercluster communicators. In del Coso et al. [16], an algorithm that considers cooperative diversity using MIMO channels in wireless sensor networks is presented, time is slotted, and one of the time slots is used for internal communication in the clusters and another time slot is used for intercluster communication. Energy constraints are also used in each of the links belonging to the end-to-end communication, and the algorithm achieves minimum outage probability in the entire trajectory by deriving the optimum times at which intra- or inter-cluster communication must take place at every hop as well as the power allocation in the route. The results show that diversity like that presented for MIMO channels is achieved using this algorithm.

5.3.5 Fundamentals of Space–Time Processing

In previous sections, we discussed paradigms of using several antennas to transmit or receive signals through wireless links. This is the principle underlying space–time processing where diversity is the fundamental idea. In Vucetic and Yuan [81], Paulraj and Papadias [65], and Paulraj et al. [66], the fundamentals of space–time processing are introduced with channel characteristics such as fading, path loss, delay spread, Doppler spread, and angle spread.

Discussion of multiple transmitters and receivers was introduced in previous sections, where channel impulse responses were also presented along with the advantages of MIMO technologies.

The received signal in the set of m elements indicated in the MIMO case was explained, and it was found that adding noise is the basis of the signal considered for space–time processing. The main problem is to determine the transmitted signal from observations of the received signal and the limited knowledge of the channel. Limitations in the knowledge of the channel are what the authors call *structure*. The objective is to characterize the spatial and temporal structure of the channel. The spatial structure contains information on array geometry, scattering, receiver gains, and so on, and the temporal structure consists of modulation format, pulse shaping, signal constellation, symbol rate, signal alphabet, and so forth.

In Paulraj and Papadias [65], an interference suppression approach is considered to apply space–time processing in order to receive the signal from a transmitter. Multiuser interference is treated as unknown additive noise. Two algorithms are introduced, basically the maximum likelihood sequence estimation (MLSE) and the minimum mean square error (MMSE). Performance of both schemes depends on co-channel interference and intersymbol interference together with channel characteristics. The authors also argue that MLSE outperforms MMSE when perfect channel estimates are available, especially for the case when intersymbol interference (ISI) is significant. In contrast, when the channel behavior has significant co-channel interference, MMSE outperforms MLSE.

Application of this technique can also be seen in switched-beam systems, and providing diversity helps to improve cellular coverage and QoS, which in turn translates into better interference suppression algorithms that increase network capacity. In principle, due to the need for accurate channel parameter estimation, networks with infrastructure will be clearly deployed in the uplink. Also, space–time techniques are signal dependent; that is, the air interface of the technology used will determine performance of the algorithm applied, and such performance will not be the same if the technology is changed.

REFERENCES

[1] J. Albowicz, A. Chen, L. Zhang, Recursive position estimation in sensor networks, Proceedings of the IEEE International Conference on Network Protocols, November (2001) 35–41.

[2] J.B. Andersen, Array gain and capacity for known random channels with multiple element arrays at both ends, IEEE Journal on Selected Areas in Communications, 18 (11) (2000) 2172–2178.

[3] S. Basagni, M. Conti, S. Giordano, I. Stojmenovic (Eds.), Mobile Ad-Hoc Networking, IEEE Press, Wiley-Interscience, 2004.

[4] R.A. Berry, E.M. Yeh, Cross-layer wireless resource allocation, IEEE Signal Processing Magazine, September (2004) 59–68.

[5] D. Bertsekas, R. Gallager, Data Networks, 2nd edition, Prentice Hall, 1992.

[6] J. Cai, D.J. Goodman, General packet radio service in GSM, IEEE Communications Magazine, 35 (10) (1997) 122–131.

[7] S. Capkun, M. Hamdi, J.P. Hubaux, GPS-free positioning in mobile ad hoc networks, Proceedings of the 34th IEEE Hawaii International Conference on System Sciences (HICSS-34), January (2001) 9008.

[8] H. Celebi, H. Arslan, Cognitive positioning systems, IEEE Transactions on Wireless Communications, 6 (12) (2007) 4475–4483.

[9] H. Celebi, H. Arslan, Utilization of location information in cognitive wireless networks, IEEE Wireless Communications, 14 (4) (2007) 6–13.

[10] S. Chakrabarti, A. Mishra, QoS issues in ad hoc wireless networks, IEEE Communications Magazine, 39 (2) (2001) 142–148.

[11] D. Chizhik, G.J. Foschini, M.J. Gans, R.A. Valenzuela, Keyholes, correlations, and capacities of multielement transmit and receive antennas, IEEE Transactions on Wireless Communications, 1 (2) (2002) 361–368.

[12] I. Chlamtac, M. Conti, J.J.N. Liu, Mobile ad hoc networking: imperatives and challenges, Ad Hoc Networks Journal, 1 (1) (2003) 13–64.

[13] D.C. Cox, Wireless network access for personal communications, IEEE Communications Magazine, 30 (12) (1992) 96–115.

[14] D.R. Cox, Renewal Theory, Methuen and Co., Ltd., 1962.

[15] S. De, C. Qiao, P. Pados, M. Chatterjee, S. Philip, An integrated cross-layer study of wireless CDMA sensor networks, IEEE Journal on Selected Areas in Communications, 22 (7) (2004) 1271–1285.

[16] A. del Coso, U. Spagnolini, C. Ibars, Cooperative distributed MIMO channels in wireless sensor networks, IEEE Journal on Selected Areas in Communications, 25 (2) (2007) 402–414.

[17] N. Devroye, P. Mitran, V. Tarokh, Achievable rates in cognitive radio channels, IEEE Transactions on Information Theory, 52 (5) (2006) 1813–1827.

[18] G. Dimic, N.D. Sidiropoulos, R. Zhang, Medium access control–Physical cross–layer Design, IEEE Signal Processing Magazine (2004) 40–50.

[19] L. Doherty, K.S.J. Pister, L.E. Ghaoui, Convex position estimation in wireless sensor networks, Proceedings of the IEEE INFOCOM 3 (2001) 1655–1663.

[20] Y. Fang, Hyper-Erlang Distribution model and its application in wireless mobile networks, Wireless Networks, 7 (2001) 211–219.

[21] K. Feher, Wireless Digital Communications: Modulation and Spread Spectrum Applications, 1st edition, Prentice Hall, 1995.

[22] M. Frodigh, P. Johansson, P. Larsson, Wireless ad hoc networking—the art of networking without a network, Ericsson Review, 4 December (2000) 248–263. Available at *www.ericsson.com/about/publications/review/2000_04/files/2000046.pdf*.

[23] J. M. Gallegos, C. Vargas, Blocking effects of outage and mobility in wireless networks, Proceedings of the IEEE Vehicular Technology Conference, 3 (1999) 2446–2450.

[24] A. Galstyan, B. Krishnamachari, K. Lerman, S. Pattem, Distributed online localization in sensor networks using a moving target, Proceedings of the IEEE Information Processing in Sensor Networks, April (2004) 61–70.

[25] A. Garca-Berumen, C. Vargas, Mobility effects in a wireless ATM network with fixed routing, Proceedings of the IEEE International Conference on Communications, 3 (1999) 1987–1991.

[26] S.G. Glisic, Advanced Wireless Communications, 2nd edition, John Wiley & Sons, 2007.

[27] A. Goldsmith, Wireless Communications, Cambridge University Press, 2005.

[28] E.G. Goodaire, M.M. Parmenter, Discrete Mathematics with Graph Theory, Prentice Hall, 1998.

[29] M. Grossglauser, D.N.C. Tse, Mobility increases the capacity of ad hoc wireless networks, IEEE/ACM Transactions on Networking, 10 (4) (2002) 477–486.

[30] P. Gupta, P.R. Kumar, The capacity of wireless networks, IEEE Transactions on Information Theory, 6 (2) (2000) 388-404.

[31] F. Gustafsson, F. Gunnarsson, Mobile positioning using wireless networks, IEEE Signal Processing Magazine, 22 (4) (2005) 41-53.

[32] S. Haykin, Digital Communications, 4th edition, John Wiley & Sons, 2001.

[33] S. Haykin, Cognitive radio: brain-empowered wireless communications, IEEE Journal on Selected Areas in Communications, 23 (2) (2005) 201-220.

[34] D. Hong, S. Rappaport, Traffic model and performance analysis for cellular mobile radio telephone systems with prioritized and nonprioritized handoff procedures, IEEE Transactions on Vehicular Technology, 35 (3) (1986) 2037-2039.

[35] Y.-W. Hong, W.-J. Huang, F.-H. Chiu, C-C.J. Kuo, Cooperative communications in resource-constrained wireless networks, IEEE Signal Processing Magazine, 24 (3) (2007) 47-57.

[36] K. Itô, H.P. McKean, Jr., Diffusion Processes and their Sample Paths, Springer-Verlag, 1974.

[37] V.M. Jovanovic, J. Gazzola, Capacity of present narrowband cellular systems: interference-limited or blocking-limited, IEEE Personal Communications, 4 (6) (1997) 42-51.

[38] S. Karlin, H.M. Taylor, A First Course In Stochastic Processes, 2nd edition, Academic Press, 1975.

[39] S. Karlin, H.M. Taylor, A Second Course In Stochastic Processes, Academic Press, 1981.

[40] F.P. Kelly, Reversibility and Stochastic Networks, John Wiley & Sons, 1987.

[41] T.J. Kwon, M. Gerla, Clustering with power control, Proceedings of the IEEE Military Communications Conference, 2 (1999) 1424-1428.

[42] J.K. Kwon, D.K. Sung, Soft handoff modeling in CDMA cellular systems, IEEE VTC '97, 3 (8) (1997) 1548-1551.

[43] A.B. Lawson, D.G.T. Denison, Spatial Cluster Modelling, Chapman & Hall/CRC, 2002.

[44] J.-Y. Le Boudec, M. Vojnovic, The random trip model: Stability, stationary regime, and perfect simulation, IEEE/ACM Transactions on Networking, 14 (6) (2006) 1153-1166.

[45] F. Legendre, V. Borrel, M. Dias, S. Fdida, Reconsidering microscopic mobility modeling for self-organization networks, IEEE Network, 20 (6) (2006) 4-12.

[46] B. Liang, Z.J. Haas, Predictive distance-based mobility management for multidimensional PCS networks, IEEE/ACM Transactions on Networking, 11 (5) (2003) 718-732.

[47] A.B. McDonald, T.F. Znati, A mobility-based framework for adaptive clustering in wireless ad hoc networks, IEEE Journal on Selected Areas in Communications, 17 (8) (1999) 1466-1487.

[48] D. McMillan, Traffic Modelling and Analysis for Cellular Mobile Networks, Proceedings of the ITC-13 (1991) 627-632.

[49] R. Meester, R. Roy, Continuum Percolation, Cambridge University Press, 1996.

[50] J. Mitola, III, Cognitive Radio: An Integrated Agent Architecture for Software Defined Radio, Doctor of Technology, Royal Institute of Technology (KTH), Stockholm, 2000.

[51] J. Mitola, III, Cognitive radio: Making software radios more personal, IEEE Personal Communications, 6 (4) (1999) 13-18.

[52] J.J. Mora-Zamorano, C. Vargas, Handoff routing strategies and mobility in multicarrier wireless networks, Proceedings of the IEEE GLOBECOM, 3, November (1998) 1402-1407.

[53] R. Miranda-Guardiola, C. Vargas-Rosales, CDMA soft handoff modeling: a networking approach, Proceedings of the IEEE Vehicular Technology Conference, 2, May (1998) 1641-1645.

[54] J. Moller, R.P. Waagepetersen, Statistical Inference and Simulation for Spatial Point Processes, Chapman & Hall/CRC, 2004.

[55] R.L. Moses, D. Krishnamurthy, R. Patterson, An auto-calibration method for unattended ground sensors, Proceedings IEEE ICASSP, 3, May (2002) 2941-2944.

[56] D. Munoz, O. Uribe, C. Vargas-Rosales, H. Maturino, Interference bounds in power controlled systems, IEEE Communications Letters, 4 (12) (2000) 398-401.

[57] S. Nanda, J. Ketchum, M. Wallace, S. Howard, A high performance MIMO OFDM wireless LAN, IEEE Communications Magazine, 43 (2) (2005) 101-109.

[58] W. Navidi, T. Camp, Stationary distributions for the random waypoint mobility model, IEEE Transactions on Mobile Computing, 3 (1) (2004) 99-108.

[59] D. Niculescu, B. Nath, Ad hoc positioning system (APS), IEEE Global Telecommunications Conference, 5 (2001) 2926-2931.

[60] D. Niculescu, B. Nath, Ad hoc positioning system (APS) using AOA, Proceedings IEEE INFOCOM, 3 (2003) 1734-1743.

[61] P. Orlik, S.S. Rappaport, A model for teletraffic performance and channel holding time characterization in wireless cellular communication with general session and dwell time distributions, IEEE Journal on Selected Areas in Communications, 16 (5) (1998) 788-803.

[62] N. Patwari, R.J. O'Dea, Y. Wang, Relative location in wireless networks, Proceedings IEEE VTC01, 2 (2001) 1149-1153.

[63] N. Patwari, J.N. Ash, S. Kyperountas, A.O. Hero III, R.L. Moses, N.S. Correal, Locating the nodes, IEEE Signal Processing Magazine, 22 (4) (2005) 54–69.

[64] A.J. Paulraj, D.A. Gore, R.U. Nabar, H. Bolcskei, An overview of MIMO communications—a key to gigabit wireless, Proceedings of the IEEE, 92 (2) (2004) 198–218.

[65] A.J. Paulraj, C.B. Papadias, Space-time processing for wireless communications, IEEE Signal Processing Magazine, 14 (6) (1997) 49–83.

[66] A.J. Paulraj, R. Nabar, D. Gore, Introduction to Space-Time Wireless Communications, Cambridge University Press, 2003.

[67] John G. Proakis, Digital Communications, 4th edition, McGraw-Hill, 2001.

[68] T.S. Rappaport, Wireless Communications, Principles and Practice, Prentice Hall, 1996.

[69] C. Savarese, J.M. Rabaey, J. Beutel, Locationing in distributed ad-hoc wireless sensor networks, Proceedings IEEE ICASSP, May (2001) 2037–2040.

[70] A. Savvides, H. Park, M.B. Srivastava, The bits and flops of the N-hop multilateration primitive for node localization problems, Proceedings of the International Workshop on Sensor Networks and Applications, September (2002) 112–121.

[71] A. Savvides, W.L. Garber, R.L. Moses, M.B. Srivastava, An analysis of error inducing parameters in multihop sensor node localization, IEEE Transactions on Mobile Computing, 4 (6) (2005) 567–577.

[72] A. Scaglione, D. Goeckel, N. Laneman, Cooperative communications in mobile ad hoc networks, IEEE Signal Processing Magazine, 23 (6) (2006) 18–29.

[73] V. Srinivasan, P. Nuggehalli, C.F. Chiaserini, R.R. Rao, An analytical approach to the study of cooperation in wireless ad hoc networks, IEEE Transactions on Wireless Communications, 4 (2) (2005) 722–733.

[74] G. Sun, J. Chen, W. Guo, K.J.R. Liu, Signal processing techniques in network-aided positioning, IEEE Signal Processing Magazine, July (2005) 12–23.

[75] Szu-Lin, Jen-Yeu, Jane Hwa-Huang. Performance analysis of soft handoff in CDMA cellular networks, IEEE Journal on Selected Areas in Communications, 14 (9) (1996) 1762–1769.

[76] J.B. Tenenbaum, V. de Silva, J.C. Langford, A global geometric framework for nonlinear dimensionality reduction, Science, 290 (2000) 2319–2323.

[77] D.J. Torrieri, Statistical theory of passive location systems, IEEE Transactions on Aerospace and Electronic Systems, 20 (2) (1984) 183–198.

[78] M. Van Der Schaar, S. Shankar, Cross-layer wireless multimedia transmission: challenges, principles, and new paradigms, IEEE Wireless Communications, 12 (4) (2005) 50–58.

[79] C. Vargas-Rosales, M.V. Hegde, M. Naraghi-Pour, Implied costs for multirate wireless networks, Wireless Networks: The Journal of Mobile Communication, Computation and Information, 10 (2004) 323–337.

[80] C. Vargas-Rosales, E. Stevens, Adaptive resource sharing in wireless networks, IEEE Communications Letters, 7 (9) (2003) 428–430.

[81] B. Vucetic, J. Yuan, Space-Time Coding, John Wiley & Sons, 2003.

[82] S.G. Wilson, Digital Modulation and Coding, 1st edition, Prentice Hall, 1996.

[83] A.H. Zahran, B. Liang, A generic framework for mobility modeling and performance analysis in next-generation heterogeneous wireless networks, IEEE Communications Magazine, 45 (9) (2007) 92–99.

[84] M.Z. Antonio, A theoretical framework for the evaluation of connectivity, robustness and reachability in wireless ad hoc networks, M.Sc. Thesis, ITESM-Monterrey, 2005.

[85] E. Baca, Performance of wireless networks using migration processes for mobility modeling, M.Sc. Thesis, ITESM-Monterrey, 2002.

[86] J.I. Bermudez, Analysis and modeling of the population distribution for cellular systems, M.Sc. Thesis, ITESM-Monterrey, 1997.

[87] P. Perala, C. Vargas, Introduction to MIMO wireless channel. Internship Technical Report, ITESM-Monterrey, May 2006.

Applications of Terrestrial-Based Location Systems

6

Position location (PL) estimation could provide added value to existing applications, as well as the creation of new ones such as emergency services, tracking and navigation, billing, and so on [5]. Positioning systems for cellular networks have been classified as those that use the global positioning system (GPS) and those that use signal strength or range to estimate location through other methods such as angle of arrival (AOA). In this chapter, we examine the PL problem within the networking area, with discussions and presentations of architectures and system points of view of different scenarios, including future technology scenarios with examples. We also relate possible solutions of the location estimation problem to scenarios and algorithms presented in previous chapters. Some technologies, such as WiFi, and ZigBee, will be addressed within the general scenarios that have been presented up to this point in the book. The sensor and ad hoc network scenario will also be presented. Tradeoffs that help to compare different technologies will be fundamental for decision-making purposes. Discussion on the use of received signal strength, time of arrival (TOA), or AOA will be put into context.

6.1 CELLULAR SYSTEMS

The cellular system can be seen as a one-hop scenario where range-based location estimation algorithms are carried out in order to provide the position of a node in terms of techniques in which the base stations play an important role by executing strategies such as those mentioned in Chapters 2 and 3. Position estimation in these systems is based on methods such as AOA, TOA, TDOA, and so on, where a centralized unit is responsible for calculating position based on noisy evidence provided by several base stations. This evidence

207

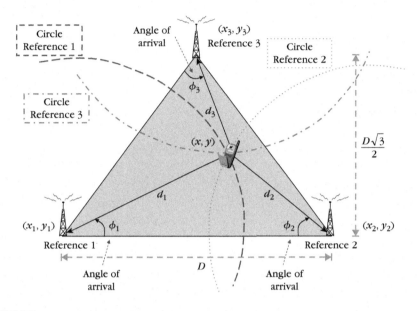

FIGURE 6.1

Example scenario of cellular system for PL estimation.

is used within a system of equations that are solved to provide the final position. The equations, sometimes, represent geometric figures (e.g., circles, ellipses, and hyperbolas). Figure 6.1, shows a scenario where circles are used to approximate a node's position. The scenario shown is not considering noise in the environment; hence the three circles intersect exactly where the node is found.

6.1.1 2G and 3G Systems

As discussed in Chapter 5, the air interface in 2G communication systems is based on TDMA and CDMA technologies as medium-access techniques. The evolution toward 3G is CDMA in two versions known as cdma2000 and WCDMA, the latter also known as UMTS. When 2G systems are considered, GSM always comes to mind due to its strong growth in terms of coverage and number of subscribers around the world. GSM incorporates GPS to locate mobiles and is known as the assisted GPS method. Because GSM requires hardware additions on both ends (mobile and base stations), it is not necessarily a method that could facilitate the needed applications. The use of AOA can have some problems, especially in urban environments where noon line-of-sight (NLOS) propagation is a strong channel impairment. TOA can be implemented in GSM by use of time advance (TA) [6], where a trip time between mobile and base can be measured. The problem with TA is that its resolution is of about 500 m. A strategy to consider is using TDOA known for GSM systems as Uplink TOA (UL-TOA) and Enhanced Observed Time Difference (E-OTD) [7–10]. Recall that TDOA provides a hyperbolic scenario to estimate position as shown in Chapter 3.

An alternative solution to the estimation problem due to channel impairments encountered in propagating the signal is the use of channel or environment signature [1], where a database of the environment stored in the network is used to compare with the received signals. The database uses measurements of path loss, power delay profile, spectral densities, and channel sounding data to obtain a picture of the conditions surrounding the base stations. This is a technique more suitable for indoor environments, especially for managing large volumes of data to be processed and stored. In Ding-Bing and Rong-Teng [2], a method for location estimation is proposed based on signal attenuation differences in the downlinks. Besides being robust to shadowing and path-loss modeling, the method provides means to implement location-based services for handsets that lack the functionality. The authors base their methodology on the use of the Hata model extended to 2 GHz (i.e., the COST231 model) and the use of a log-normal random variable for shadowing. The model is used to determine the path loss difference between two base stations. Such a difference is then associated to circles around the base stations that intersect at several points depending on the positions of base stations and mobiles. Normally, these intersections are defined as points and an average is calculated to estimate the position, but in Ahonen and Laitinen [1] a centroid is considered the desired position.

Results are introduced where environment variation is considered through use of the standard deviation of the log-normal random variable used for shadowing. It is shown [1] that for low shadow fading with a standard deviation of 2 dB the number of base stations needed in order to have an error of less than 40 m 67% of the time is five. The authors conclude that their method outperforms the cell-ID method in a GSM network in Taipei; they also discuss the possibility of applying such a method to a UTRAN network by modifying some parameters in the method.

A review of location methods for CDMA as a 2G technology is found in Caffery and Stuber [4]. These techniques are based on traditional radiolocation methods that base their decisions on signal strength, TDOA, TOA, and AOA. In addition, algorithms are introduced for the case of TDOA and TOA [4], with discussions on the main problems caused by the multipath of the signal propagation—the multiple access interference (MAI) and the non-LOS signal conditions. The conclusion is that TOA or TDOA is the choice when accuracy is a concern in locating a mobile; otherwise, signal strength or network organization in terms of location areas or cells and sectors might be used.

The efforts to standardize positioning are discussed in Zhao [9], where the focus is on techniques for 3G technology. In addition, techniques are based on AOA, TOA, or TDOA as seen in previous chapters. The discussion is concentrated on the location technologies for 3G specified by 3GPP, where three options are considered—that is, cell-ID, observed TDOA (OTDOA), and assisted GPS (A-GPS). The OTDOA method is based on TDOA, where hyperbolas are formed by the reference base stations and location is determined by the trilateration. It is also known that the more references we have, the better the accuracy. Base stations' location must be known either by the network or by the mobile to be located. Location measurement units (LMUs) are used since these systems have greater capabilities than a mobile station but are less complex than a base station. LMUs are deployed at known locations and provide timing information from various points in the network coverage area; data are collected from local transmitters. This information is processed to obtain relative time difference, and sent afterward toward the base station of the radio network controller or to the mobile station to be located. Problems include the lack of synchronicity among the base stations, especially for FDD mode; location of base stations to obtain improved multilateration techniques; and capacity loss due to the use of such a location service.

Standards for advanced forward link trilateration (A-FLT) and A-GPS are efforts carried out by Telecommunications Industry Associates (TIA) through the IS-801 standard. In A-FLT, time difference or phase delay is obtained from the CDMA pilot signals of the base stations. These differences are obtained from the pilot signal of the base station serving the mobile to be located and those of neighboring base stations [8]. A-FLT has an accuracy of 50 to 200 m and a 1/8-chip resolution, uses control messaging through IS-801, requires handset software changes, and is generally used in a hybrid form with A-GPS. Table 6.1 summarizes

Table 6.1 Location Estimation Techniques in Cellular Systems

Characteristics		
Method	**Technology**	**Accuracy**
Cell-ID	All	100 m–3 Km
Cell-ID with TA	GSM	500 m
AFLT	CDMA	50–200 m
EOTD	GSM	50–200 m
AGPS	All	5–30 m
TDOA	All	100–200 m
AOA	All	100–200 m

the technologies for location estimation in a cellular environment, the type of network technology where the estimation method can be implemented, and the approximate accuracy of the estimation.

6.1.2 Multihop Cellular

The main idea behind the use of multihop routes to achieve connectivity can be seen in Figure 6.2, where in a cellular scenario a mobile is using other similar subscriber equipment to achieve connectivity to the infrastructure. Note that even if such a figure shows a situation where a mobile is assisted in the communication link toward the base station that is supposed to serve it, the multihop could very well be performed to a base station where the node in question is not within its coverage area (see Le and Hossain [11]).

Multihop cellular networks have several important issues to address, including for example resource allocation in performing multihop scenarios without the adverse effects of co-channel interference to the same serving cell or adjacent cells. This frequency planning at the multihop level would increase signal efficiency and quality, as well as coverage and capacity. Another advantage is signal propagation since path loss will be diminished due to the use of shorter links. In addition, the development of collaborative and cooperative algorithms is underway for diverse tasks such as processing, transmission, and location estimation. Multihop architectures have been proposed using OFDM [18] and the standard IEEE 802.16 [19], which specifies a network mechanism to establish links with neighbors.

The use of multihop can also be extended to networks that provide the wireless local loop (WLL) to telephone subscribers. This scenario would be similar to that of a cellular network with mobile users. The main purpose could be to save transmission power while increasing coverage. To provide localization,

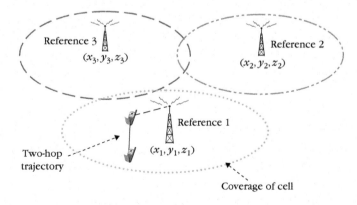

FIGURE 6.2

Example of a multihop scenario for cellular systems.

cell-ID could be used to locate the mobile in low-accuracy mode; then the multihop scenario could be used to locate the mobile with better accuracy via algorithms such as those introduced in Chapter 4. The cell-ID would basically not require any modification in the base stations or the handsets; the multihop location would require software changes to handsets that could be implemented online within the cognitive radio paradigm.

6.1.3 Cell-ID

Cell-ID is based on wireless cellular network organization, where the information on the sector of a cell at which a mobile has registered lately is used. This method has been recommended by 3GPP. During an active call, the normal resources during a communications session are used to establish the sector at which the mobile is found. The main problem with this method is accuracy, since the size of the sector plays a major role in the area to be searched. Improvements to this method can be found in Borkowski et al. [3] where the method is combined with parameters such as round-trip time (RTT).

Information about the area in which the mobile is found can be obtained from the location area update, or by paging or cell update messages. This method is always operational, so even in the case of a nonexisting location technique, cell-ID can be implemented to provide a low-accuracy location. The drawback of the method is that it does not necessarily satisfy accuracy requirements since it depends on the size of the area where the mobile is found.

In general, we can conclude that cell-ID has the benefits of low cost, no need for upgrades, and privacy, and it can be implemented with current technology. Disadvantages are low accuracy, and effectiveness is also low since it does not improve with proximity. Another disadvantage would be that the system generally assumes the mobile to be located is connected to the closest base station, but this may not be true in cases of severe multipath propagation, where transmission power from another BS can change the cell assignment to that of the cell with the best signal received. One of the best applications for cell-ID is location-based services (LBS) such as the search for places like restaurants or hospitals in a given area surrounding the mobile.

6.1.4 E911

Regarding the E911 service, the first thing that comes to mind is the FCC set of requirements that need to be satisfied to provide such service. The requirements depend on the type of solution; for instance, the network-based solution has a requirement of 100 m for 65% of the time and 300 m for 95%. For location estimation solutions based on the mobile station, the service provider must satisfy an accuracy of 50 m for 67% of the time and 150 m for 95%.

E911 stands for enhanced 911 service, which is an automated system that works together with service providers and carriers to determine user location via their telephone number. The current status of the service implemented is Phase II. The basic operation of E911 service follows. First, a call is received at a number that is defined as universal (e.g., 911); then the network uses such numbers to find a route to an emergency response center where the call is received and caller identification begins. After this procedure, in the case of a conventional call, the service can obtain the address of the caller by searching a database; however, in the case of a wireless caller, a location estimation procedure must be ordered from the service provider to transmit the position of the mobile.

6.2 LOCAL (INDOOR) NETWORK SCENARIO

The frequency bands used in the indoor scenario are higher than those of cellular scenarios. This change to a higher frequency affects propagation negatively since signal strength is inversely proportional to at least the square of the frequency [12].

In basic terms, the location problem for an indoor scenario can be seen as a well-defined area in which a node or a tag needs to be located. The space in which location estimation takes place changes if objects or nodes change with mobility. Normally, any algorithm in this scenario is benefited by processing of parameters or variables that describe the environment in which the estimation is taking place to adapt the procedure to estimate a more accurate position. Some of those parameters are not directly obtained and need to be estimated, especially those related to channel impairments.

In general, location-aware applications at the indoor level include emergency services, product marketing, navigation and guidance, tracking, and so on. The location process is not directly benefited by the use of GPS since in high population density areas and at the indoor level it is unreliable because the line-of-sight (LOS) signal requirement is not always present; this can become a problem of inaccuracy for PL estimation.

Generally, the techniques used for position location at the indoor level are basically the same as those for outdoor or cellular scenarios. The techniques are based on multilateration or triangulation such as those seen in Chapters 2 and 3. All these methods are usually based on range measurements (e.g., TOA, TDOA, AOA, DOA), which estimate distance from a base station or access point to a node of interest or tag. The basic idea is to use *proximity*, which is nothing more than recognizing the coverage area in which the node in question is located. This proximity idea increases accuracy as the number of base stations increases.

Another technique used is the channel sounding at the outdoor level or signal signature at the indoor level, better known in some areas as *fingerprinting*; in other words, a *map* of the coverage area of the network is obtained by measuring received signal strength (RSS) at various points and using a propagation model for the entire area. After this, when a node needs to be localized, the RSS from it can help locate the area in which it is found with certain accuracy by comparing the current RSS levels with those stored from fingerprinting procedures.

The main problems associated with these techniques at the indoor level are related to signal propagation. For example, more obstacles will be present in this scenario; thus, shadowing can get to critical levels producing high power loss in the signal. Also, the presence of multipath is severe since the space covered is relatively small; antennas are reached by far more significant power levels of signal components such that fading might be increased.

6.2.1 Technologies and Standards Review

Of technologies that can be used for location estimation, we can mention WiFi, Bluetooth, and radio frequency ID (RFID). WiFi technology, better known as the corresponding standard IEEE 802.11*x*, is used mainly as a WLAN where users receive data applications such as Internet access. A WiFi network is mainly organized by an infrastructure consisting of access points (APs) that are interconnected to the Internet by means of routers or switches. Each AP provides a coverage area with a range of 50 to 100 m using various channels in the frequency band of 2.5 GHz or 5 GHz. The medium access control (MAC) used is based on the carrier sense multiple access protocol with collision avoidance (CSMA/CA), where a node asks for permission to use the channel, and when the channel is granted no other station within that coverage area can use it. The permission is subject to collisions, but the transmission of information is not. WiFi generally requires LOS, but it has certain robustness at mild levels of shadowing. As the number of nodes increases within the coverage area of an AP, the throughput decreases with levels of 5 Mbps for 11-Mbps channels.

Another technology is the RFID, widely used for logistic purposes and inventory services. It does not require LOS, and the network formed by RFID nodes can have several readers or nodes with more intelligence and identification nodes or tags. Tags are basically of two kinds—passive and active. The difference between the tag types is in the range that can be read, cost, and lifetime. Proximity is well suited for location estimation in these networks, and tag density also affects accuracy. Bluetooth is another standardized technology where low-range personal communications are carried out. It consists of devices of low cost and low bandwidth with ranges normally in the 10- to 20-m range. It uses spread spectrum in the frequency hopping (FH) case.

Other technologies include infrared, which is generally expensive compared to those just described. Infrared also has low coverage and low power, and it is adversely affected by sunlight. Coverage is about 10 m with proximity as the basic scenario for location estimation. The LOS requirement is also strict in order to establish links; thus, it would be difficult to apply any of the multihop methods described in Chapter 4. Ultrasound, another technique used in the indoor scenario, is low cost, fairly precise in the measurement of parameters especially for high-density scenarios, and requires LOS to communicate.

6.2.2 Localization with WiFi, Bluetooth, and ZigBee

As mentioned previously, WiFi refers to a WLAN technology, the main application of which is access to data services. The fundamental architecture of WiFi systems is shown in Figure 6.3, where APs are used to interconnect wireless users to the infrastructure that provides the data services.

Selected technologies for indoor location systems are MIT Cricket, Ekahau, Intel Place Lab, Microsoft RADAR, Skyhook Wireless WPS, BLIP Systems, and

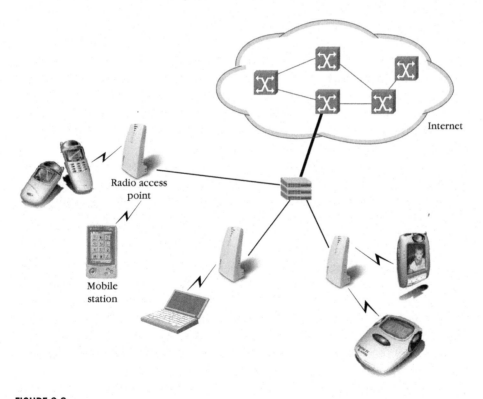

FIGURE 6.3

Basic WiFi connectivity scenario.

Aeroscout [16]. For example, Cricket is based on deployment of AP or beacons on ceilings and walls. Nodes can produce location information when they receive an RF signal from the beacons emitting an ultrasonic signal moments later. With this ultrasonic signal, nodes can estimate range from the beacons. The receivers use an estimation algorithm based on maximum likelihood to decide which beacon is closer. This system uses passive nodes listening to active beacons transmitting RF and ultrasonic signals.

In the WiFi system, location can be determined using several techniques, including RADAR by Microsoft Research, which provides location estimation with an accuracy of 3 m [14]. The RADAR system is based on the concept of signal signature or fingerprinting where RSS is measured on a grid in the area of interest, identifying the APs that are received at each of the points of the grid with a 1-m separation for each point of the grid. The major problems with this system are its scalability and calibration, as well as channel impairments that produce a time-varying response that changes the point of view that the fingerprinting data provide. Another system, known as DSSS RADAR, it is made up of WiFi in a PCMCIA card at 2.4 GHz with a transmission rate of 11 Mbps. Since the signal is Direct Sequence Spread Spectrum (DSSS), the transmission bandwidth is 20 MHz, using Barker codes for spreading [13]; the system is used together with MIMO. The system determines targets, such as the position of APs, using RSS and provides parameters, such as elevation. Some limitations of RADAR are that every node to be located needs to have a WiFi network interface which is not necessarily small for some devices. Also, the accuracy is not appropriate when the node to be located is on another floor of the building. Another problem is the time-varying nature of the channel, which is not necessarily captured in the fingerprinting maps.

Another option within WiFi is Place Lab [15], which uses an interface such as that in RADAR, but with an advantage that fingerprinting is not needed. The device bases the location estimation on the detected APs' positions. Infrastructure APs have known locations that are fed to the wireless devices through a small database. Also, when large areas need to be covered, Place Lab works with GSM networks and Bluetooth devices. For large areas, GSM is used, while for small areas and to increase accuracy, Bluetooth devices are used. Since Place Lab needs almost no calibration, it has less accuracy than RADAR, where 20 to 25 m of accuracy is achieved in urban areas with moderate density. APs transmit a signal with unique identifiers that are recognizable by the nodes using Place Lab, and the identifier is matched to a position in a database.

Aeroscout is a WiFi positioning system that is able to locate computers, bar code scanners, laptops, RFID readers, and basically any node that has an IEEE 802.11b and 802.11g standard card. The technique is based on triangulation using TDOA and RSS [16]. When the area to be covered is large, TDOA is preferred; in contrast, RSS is used for small areas. Aeroscout uses tags that can be controlled through the WiFi network. These tags together with location receivers are used to generate a TDOA technique to locate a node via triangulation.

Bluetooth systems include BLIPNet and BLIP [16]. Each system offers access to a local network through the Bluetooth technology. The system consists of Bluetooth devices, applications, BLIPNodes, and a server. BLIPNodes are capable of registering RSS values. The system relies on a positioning module in which several BLIPNodes are deployed in the area, and when location is necessary, tags must be within 100 m of a BLIPNode. Multilateration is performed by the positioning module with the BLIPNodes that receive the signal from the device to be located. In general, problems arise with all of these systems; for example, installation and deployment require calibration, special hardware, tags, beacons, and previously decided APs' locations to serve as anchors; operation and maintenance can be an issue, as well as cost and scalability of the system and recalibration.

ZigBee is another technology for wireless communications similar to WiFi. Its standard is IEEE 802.15.4. The point of view of the location estimation problem for ZigBee is the context-aware system scenario where applications react and adapt to sensed data and the environment; for example, it could react to temperature or pressure as environment variables sensed, or it could react to a calendar of activities such as network maintenance or scheduling of network tasks. The location estimation information is also within these context-aware applications. One example of a system combines the advantages of Place Lab, where no fingerprinting is necessary, with the use of ZigBee for transmission of information. The tags could use multilateration with ZigBee signals through some sensors, or ZigBee nodes could be used as beacons with extra features that allow the system to locate nodes.

One of the advantages of using ZigBee for location is that sensors can be involved; thus various kinds of parameters can be used for location estimation. For example, besides using RSS, nodes could use extra sensors such as gyroscopes, accelerometers, or sometimes a GPS to perform location estimation using ZigBee nodes. Location monitoring can also be implemented using ZigBee by dividing the nodes by their characteristics. For instance, some nodes can be fixed or static with known positions, interconnected forming a network, and a node can be working as a gateway to which the rest of the nodes can connect either directly or in a multihop fashion. The problem would be to locate mobile nodes. Static ZigBee nodes can be used as APs and the scenario could be translated to any of the scenarios presented in Chapter 4, where heuristics can be used to estimate position. The main part of the estimation in this multihop fashion is the use of a routing algorithm that could provide information not only of the next hop but also of the entire route connecting origin and destination. Also, accuracy can be increased as the node density is increased, but there is a trade-off with computational timing and computational complexity.

We can locate static ZigBee nodes so that the entire area is covered with a connected network topology, and any other node can be connected to three or more of these static nodes at any time. In an indoor scenario, coverage of a node is regularly 20 m due to multipath propagation. If accuracy needs to be improved, fingerprinting of the environment can be used by mapping the area when the static nodes are active and registering this information in a database to be used by the system to locate mobile nodes in the network.

6.2.3 RFID and INS

Radio-frequency Identification, best known as RFID, consists of devices that can be attached to several objects in order to serve as identifiers. The areas holding the objects have readers installed that can detect the presence of such devices. These readers generally are fixed and have greater processing power. In addition, the devices can be fixed and help to locate a moving reader. In security situations, such as firefighting, they can serve as the readers that would need to be located without using GPS. When a node must be located in an indoor environment, the system that performs such tasks is called the indoor navigation system (INS). The most widely used navigation system is GPS, but recall that at the indoor level GPS has performance problems due to multipath propagation that directly affects accuracy. A navigation method that could be used with RFID technology is dead-reckoning, if sensors suitable for parameter measurement are available.

To determine location using RFID, the channel or propagation environment needs to be assessed, especially for emergency scenarios. The network of RFID devices could include sensors such as accelerometers and pressure or light sensors. When RFID devices are considered, some additional issues need to be addressed such as interoperability, cost, and management. The basic RFID system consists of RFID devices or tags that provide the identification of objects and readers that communicate with the tags and with a set of servers that provide the appropriate identification information from a database. The readers send an RF signal within their coverage area and tags are able to recognize such a signal and send back information about themselves. There are passive tags that have no integrated energy source, and the energy to transmit back the information is derived from the RF signal received. The radio range is limited. Tags can also be semipassive, meaning that they have an energy source and the only range limit derives from reader sensitivity. Tags can be active, which means that they have an integrated power source and can transmit required information or other types of information that could be programmed. In the United States, the RF channels used are 500 kHz of bandwidth in the 900-MHz band, and the maximum transmission power is 4 W. In contrast, in Europe the channels are of 200 kHz in the 860-MHz band transmitting at 2 W. Reader sensitivity is around −90 dBm (see Krishna and Hasak [17]).

Location estimation can be achieved by comparing the network to the scenario for WLANs where one has tags communicating with APs (readers) that could have fixed and known positions. Then algorithms such as those presented in Chapter 4 can be used in a multihop scenario to locate tags based on multilateration via readers.

6.2.4 System Comparison

Quantitative and qualitative comparisons among various location estimation techniques, some for outdoor and some for indoor, are provided by several researchers [39–41]. In Sun et al. [40], a comparison of standard and nonstandard positioning for outdoor environments is presented, where the focus is on the extra system components required to perform localization. For example, using E-OTD for GSM requires a change in software for accuracy of 50 to 100 m, but response time is generally slow. For CDMA using AGPS, accuracy of 10 m or less is possible, but every handset must have AGPS capability and reception. With WCDMA using OTDOA, changes in the base station are necessary for accuracy of less than 50 m using trilateration or multilateration. In contrast, as mentioned earlier, cell-ID needs no change in software or air interface and accuracy depends on area size. Accuracy can be improved if used in a hybrid way with other methods. On the other hand, Sun et al. [40] also introduce nonstandard methods where algorithms (e.g., pattern matching) can be used that require a change of hardware and certainly of software. Also, smart antennas are introduced as nonstandard methods where modification of the base stations is needed but not of the handsets.

In contrast, in Gustafsson and Gunnarsson [41] techniques from the point of view of the measured parameters are compared, providing accuracy in terms of the obtainable standard deviation of the measurement error. In Porretta et al. [39], a similar comparison is presented with an extra parameter—mean value of the location error—and a discussion on what is needed to estimate position is included. For example, for TOA three base stations are needed to perform triangulation, with a mean error of 25 m and a standard deviation of 14 m. Using an array antenna at the base station and a fingerprinting database provides accuracy of less than 25 m for 67% of the time.

6.2.5 System Trade-offs

In general, system trade-offs depend widely on environment parameters and system features. For example, transmission power depends on remaining battery energy, which affects coverage and determines connectivity so that other nodes can be reached and eventually located. To prevent such problems, compromises must be made to ensure all network processes and tasks are carried out in the long run. Use of RSS is affected by statistics and data collection; for example, manufacturer and user proximity affect environment awareness, as well as the period of time used for the measurement, the interference caused by other nodes, and the physical characteristics of the area where the network is operating. Even measurements made by the nodes may differ according to manufacturer since receiver sensitivity may vary.

When one has the possibility of increasing node density, interference levels can be affected, but location accuracy may improve since more references can be used to perform multilateration and internode distances can be estimated with greater accuracy as they are shorter. The cost of devices will depend on hardware features that could improve location estimation, and this also can place constraints on accuracy. Higher costs could provide devices with GPS features that in the outdoor scenario could be advantageous, but in indoor scenarios spending money for better sensors might be the rule to improve accuracy.

6.3 MESH SYSTEMS

Wireless mesh networks are a new option for the last mile. They can provide Internet access with broadband services, in addition to integrating technologies and tasks as diverse as a 3G and WiFi or a voice service and collaborative data mining with smart sensors. Each of the nodes in the network can work as an end user terminal or as a router cooperating in the delivery of information between an origin and a destination. The same nodes could adapt themselves to interact accordingly to changes in the environment and mobility so that they create a network topology that is maintained to provide an infrastructure for end users. Nodes could have different levels of intelligence, as well as various features to carry out communication tasks such as topology establishment, topology control, routing, scheduling, multiple access, sensing, and coding [24]. One of the advantages of a mesh network is that it can grow one node at a time; that is, nodes are capable of adapting the topology online so that new nodes can be added without any configuration. It is expected that nodes have sufficient intelligence to communicate with their neighbors and establish themselves within the topology so that all communications continue transparently. Nodes can be fixed or mobile, and could have specific tasks as well such as information sensing and processing. Other nodes could work as gateways or controllers. Next, some nodes could be deployed in an area so that connectivity is guaranteed regardless of the density of nodes present. For instance, nodes could be monitoring in a passive way so that when needed they become active to provide a coverage area for network users.

Some technologies, such as WiFi and Bluetooth, are being studied so that mesh networking is possible with them [23]. Peer-to-peer communication possibilities with those technologies are common, but their nodes do not have router capability. Also, RFID has been proposed to be integrated with sensors so that a mesh network could be formed [22]. Multimedia delivered by sensor networks to end users has been considered, (see Akyildiz et al. [21]). In mesh networks, sensors can be used to obtain better accuracy in location estimation [20]. For example, sensors such as gyroscopes, compasses, altimeters, and accelerometers can help determine position by providing variables such as velocity, height, and direction. Other sensors, such as light and temperature, can help determine the area in which a node is located by comparing the levels read out with those maintained in a database obtained by fingerprinting; in this case, not only is a signal signature provided by RSS but also a signature of variables such as luminosity and heat levels. Other sensors could be used to capture images (e.g., Webcam, CCTV) and then those images are compared to a set of images stored previously.

Mesh networks also depend on air interface standards, so a mesh network based on IEEE 802.16 WiMAX, or using WiFi or even ZigBee, could be deployed. One of the most important problems affecting communications of any kind in these networks is interference; thus, analysis and modeling of interference to determine network capacity must be carried out.

6.3.1 Sensor Networks

Sensor networks are a subset of mesh networks where nodes have the special capability of sensing some parameter such as light intensity, temperature, pressure, and velocity. In general, nodes will have limited computational and processing capabilities due to the sensing tasks. The nodes also will be able to communicate to other nodes using certain protocols. Some sensors can have greater capabilities and be ready to function as routers as well, but this is not the rule. There will be nodes in charge of collecting and processing information to be transmitted later to a central node that could be a decision-making element if the sensing involves processes such as those in power or petrochemical plants. At the indoor level, navigation sensors could be used in order to measure parameters that can be communicated to a processing unit in order to provide location estimation. Sensors, such as pressure and compass, could be used to detect direction or orientation, whereas accelerometers provide parameters from which distance can be computed.

A sensor network could be constructed of nodes working with a communication technology already standardized such as ZigBee or WiFi. RFID tags can sense certain variables, or motes (low-power microcontroller chips) could communicate in the same form. With sensors like these, some nodes with greater processing capabilities could run calculations and transmit them using another technology. For instance, RFID tags communicating to an RFID reader could use ZigBee to transmit the information to another node, and this could be connected to an AP that serves as the gateway to the Internet.

Sensor networks have been studied extensively in the literature from the perspective of the network layer, with algorithms and procedures proposed to implement neighbor discovery, topology establishment, clustering, collaborative processing, and so on. Location estimation algorithms have also been proposed. One kind that could be helpful is the algorithm that uses relational information such as that discussed in Chapter 4, where nodes use information of neighboring nodes by the RSS parameter to estimate separation distance to create a picture of the actual topology. By using anchor nodes, or APs with known positions, location estimation can be implemented. Another work using relational information is Patwari et al. [37], where the feature of self-awareness is needed in the nodes to construct and configure a topology. RSS and TOA are used in an indoor environment, and accuracy of 1 and 2 m is achieved with TOA and RSS, respectively.

Another strategy to estimate position can be implemented using the idea of local maps, where a node to be located is found using routing from a node with a known position [34]. The nodes along the route followed can construct a map of their local neighborhood with relative positions, and then report it to the node originating the location procedure. This originating node can process the information and perform transformations to align the various maps received and eventually translate the local coordinates of the maps to a coordinate seen from the originating node or a global coordinate if available. This method uses the concept of divide and conquer, by providing means to perform location at longer distances with the advantage of working with local views of the topology. This is important since connectivity is not a global- or network-level parameter. The connectivity that a node experiences depends on its local neighborhood, and even though a node may be connected to several neighbors, that group of nodes can be disconnected from the general network and not know it for some time.

In some networks, information to be sensed and processed is critical and needs to be secured. In addition, a network can better protect itself if location is performed so that intruders not only have a name but also a relative position in the network. Security procedures can be carried out in a local way too, depending on critical issues. In Ekici et al. [35], a packet traversing several nodes along a route is verified by location estimation and separation distance between origin and destination. This idea could prevent intruders from changing critical control information such as routing tables in certain regions where for example black hole attacks can be performed. By verifying this information, nodes could know whether the messages are authentic. In Zou and Chakrabarty [36], the authors consider another important problem in this kind of network where nodes are capable of carrying out tasks by themselves, consuming power from a source that is limited. Algorithms could be improved in this direction by limiting tasks such as sensing, transmission, or processing to the amount of remaining energy that the node has. In Zou and Chakrabarty [36], the network is organized in clusters where each cluster has a node with greater processing capabilities working as a cluster head. The cluster head is in charge of deciding whether a localization algorithm must be performed to find the sensors that need to provide more information for processing. In this way, a thorough on-demand sensing task is performed according to the cluster head decision.

Applications for sensor networks include surveillance and habitat and environment monitoring. It is also known that sensor nodes must be able to adapt to very different and changing operation conditions in order to provide reliable services, with limited processing, communication, and computation resources. Cost is another limitation, since sensors will not be able to have as much hardware as needed to perform, for example, communications from known positions with an LOS link. A network with low-cost sensors may consist of sensors with different localization algorithms that will be executed according to a decision-making process that depends on environment parameters (see Stoleru et al. [38]). The authors propose to use a localization manager that could work on the decision to use a specific localization algorithm depending on the conditions experienced.

One of the ideas discussed in [38] is that protocols can interact in several ways; because they are not organized as a vertical stack, the only option is as in the classical models. They could have a hierarchical structure but also a nonvertical dependency. A framework using the hierarchical protocol approach was implemented. Simulations were performed in a square area, with sensors having a transmission range of 15 m. Another important supposition is that 10% of the sensors have a GPS device. The square area had obstacles to limit the establishment of LOS links, even with satellite links affecting those sensors with GPS as well. Several localization methods, such as DV-hop, robust, centroid, and spotlight, were compared. The robust algorithm achieved accuracy of less than 3 m more than 95% of the time.

6.3.2 Ad Hoc Networks

Mobile ad hoc networks (MANETs) are wireless networks consisting of a collection of wireless mobile nodes interconnected by multihop communication paths. These nodes can freely create and dynamically self-organize an arbitrary and temporary wireless network among themselves, allowing people and devices to seamlessly internetwork in areas with no pre-existing communication infrastructure or administrative support, unlike conventional wireless fixed networks [25]. The ad hoc networking concept is not a new one, having been around in various forms for more than 20 years (e.g., in military and police areas), especially under disorganized, hostile, tactical, rescue, and disaster recovery conditions. Recently, home or small office networking and collaborative and mobile computing with laptop computers in a small area (e.g., a conference room or classroom, a single building or a convention center) have emerged as other major areas of potential applications [26].

Most of the well-known complications in mobile radio systems originate directly or indirectly from the topology instability caused by motion [27]. The topology of ad hoc networks changes dynamically as mobile nodes join or depart the network or radio links between nodes become unusable. The constant independent moving of the nodes causes the frequent failure and activation of the links leading to increased network congestion being a potential source of service impairments in MANETs, degrading the QoS seen by users of the network [26, 28]. Performance metrics, such as connectivity, robustness, fragility, reachability, throughput, and others, permit evaluation of network performance [29]. A problem related to node population and node mobility is the connectivity, which can be low when nodes are organized in different groups and these groups are isolated; consequently, designing a theoretical framework for the study of the parameters mentioned before using a more stable network topology is necessary [30].

QoS is a broadly used term referring to network capabilities that result in user satisfaction. As broad as this definition is, so too are proposals on how to deliver QoS in ad hoc networks [31]. It is important to mention that we are not making QoS, but rather we can build a model to evaluate QoS. For the proposed model, we need nodes with previous knowledge of their connectivity for functions such as routing. It is assumed that a process of discovery of network nodes exists, where one node gets the knowledge of other nodes from part of the network. The discovery of a node does not imply its reachability since nodes are mobile and links might be created and broken. Second, a routing algorithm is assumed to work on the topology of the network formed by the links established by the nodes within their coverage regions. In this form, routing provides the set of paths available in the network, and it is precisely this information that indicates the cost of the links and thus the reachability of other nodes. With this, nodes can obtain information on connectivity and robustness, which at the same time can be used to acquire a network QoS measure. Therefore, a connected node implies a reachable node, and this fact implies as well a discovered node. This grade of implication is summarized in Figure 6.4. This model of implication is related to the connectivity problem discussed in Section 5.1.4, and it is further developed in Antonio [42].

This implication model presents a new concept of network metrics that should be considered as part of performance evaluation models. With this implication model an analytical method for evaluating a performance metric (e.g., connectivity) is developed. This analytical model could be extended as a basis for the calculation of other performance metrics (e.g., reachability, robustness, and QoS). With the proposed model, we calculate the node's connectivity and this analysis could be used to imply a measure of node reachability as well as to calculate a measure of robustness.

Location estimation in these networks has some advantages with respect to sensor networks since in general all nodes in the network have higher computational power than sensor nodes and can communicate and route information; that is, routing tables and neighbor lists together with topology maintenance algorithms are executed by the nodes as a set of common tasks to be performed during operation. The algorithms in Chapter 4 with respect to heuristics with multihop scenarios can be implemented in ad hoc networks without worrying about complexity. The major disadvantage is the battery life of the nodes that now have consumption derived from the transmission and processing of

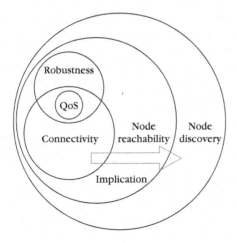

FIGURE 6.4

Model of implication of performance metrics.

information not necessarily destined for the node carrying out the processing and transmission. Position estimation in these networks can be performed using one-hop techniques for measurement of parameters in principle, such as TOA, TDOA, AOA, and carrying out algorithms such as APS [32, 33]. In order to provide location information, it is not necessary that all nodes be capable of measuring parameters such as AOA. Only some of the nodes could have such a feature and provide location information on all nodes in the network by using APS [33]. Deployment of ad hoc networks must be easy; that is, manufacturers must develop nodes capable of creating networks by themselves and providing connectivity services to end users with applications such as location-based services. Node density helps to obtain better accuracy in the position estimation, since more nodes could work as references or beacons when multilateration is performed.

6.3.3 Natural and Human-Made Disasters

There are scenarios where human-made disasters such as fires can be candidates for deployment of a wireless sensor network. For example, each firefighter can have a device that could facilitate communications with other groups to coordinate rescue missions. These devices could be capable of communication, routing, processing, and location estimation. The devices could be equipped with sensors that could provide information on concentration levels of gases that may cause an explosion or are harmful in other ways.

Natural disasters are also opportunity areas where ad hoc or sensor networks could be deployed even by nontripulated, or remote-controlled, vehicles that could set sensors in various places to establish a network. For example, after a hurricane or an earthquake, military teams can deploy nodes such as sensors, and communications can be established from the exterior to the center of the disaster area, even for broadband services. Sensors could be used to mark points in the damaged area where medical services, food services, or security teams are needed, and these points could be located by any of the algorithms introduced in previous chapters.

As one can see, nodes and sensors for these kinds of services require special features such as robust hardware in order to withstand rain or flooding impact. They should also be equipped with longtime-operation batteries, and in some cases nodes should adapt so that in determining a topology and a coverage of the network, one or more might turn themselves off and on as needed by the network. Since these situations are emergency scenarios, some of those nodes can be equipped with GPS systems so that positioning can be achieved using them as references or beacons. The rest would be communicating nodes and sensing nodes. RFID tags could also be used to mark objects that require attention. For example, a person could be at risk in disaster recovery conditions, then RFID tags could be used to detect movement and activate sound alarms to alert that person of the danger in the area.

REFERENCES

[1] S. Ahonen, H. Laitinen, Database correlation method for UMTS location, IEEE Vehicular Technology Conference, VTC03, 4, April (2003) 2696–2700.

[2] L. Ding-Bing, J. Rong-Teng, Mobile location estimation based on differences of signal attenuations for GSM systems, IEEE Transactions on Vehicular Technology, 54 (4) (2005) 1447–1453.

[3] J. Borkowski, J. Niemela, J. Lempiainen, Performance of cell ID+RTT hybrid positioning method for UMTS radio networks, Proceedings of the 5th European Wireless Conference, February (2004).

[4] J.J. Caffery, Jr., G.L. Stuber, Overview of radiolocation in CDMA cellular systems, IEEE Communications Magazine, 36 (4) (1998) 38–45.

[5] C. Drane, M. MacNaughtan, C. Scott, Positioning GSM telephones, IEEE Communications Magazine, 36 (4) (1998) 46–59.

[6] M. Pent, M.A. Spirito, E. Turco, Method for positioning GSM mobile stations using absolute time delay measurements, Electronic Letters, 33, November (1997) 2019–2020.

[7] M.A. Spirito, M.P. Wylie-Green, Mobile stations location in future TDMA mobile communication systems, IEEE Vehicular Technology Conference, 2, September (1999) 790–794.

[8] Broadcast Network Assistance for Enhanced Observed Time Difference (E-OTD) and Global Positioning System (GPS) Positioning Methods, 3GPPTS 04.35, Release 1999, v8.3.0 in January 2001. Available: *http://www.3gpp.org./ftp/Specs/html-info/0435.htm*.

[9] Y. Zhao, Standardization of mobile phone positioning for 3G systems, IEEE Communications Magazine, 40 (7) (2002) 108–116.

[10] M. A. Spirito, On the accuracy of cellular mobile station location estimation, IEEE Transactions on Vehicular Technology, 50, May (2001) 3674–3685.

[11] L. Le, E. Hossain, Multihop cellular networks: potential gains, research challenges, and a resource allocation framework, IEEE Communications Magazine, 45 (9) (2007) 66–73.

[12] T.S. Rappaport, Wireless Communications, Principles and Practice, Prentice Hall, 1996.

[13] N.B. Sinha, Measurement of target parameters using the DSSS RADAR, Progress in Electromagnetics Research, 1 (2008) 185–195.

[14] J. Hightower, G. Borriello, Location systems for ubiquitous computing, IEEE Computer, 34 (8) (2001).

[15] B. Schilit, A. La Marca, G. Borriello, W.G. Griswold, D. McDonald, et al., Ubiquitous location-aware computing and the place lab initiative, 1st ACM International Workshop on Wireless Mobile Applications and Services on WLAN Hotspots, September 2003.

[16] K.W. Kolodziej, J. Hjelm, Local Positioning Systems, LBS Applications and Services, CRC Press/Taylor & Francis, 2006.

[17] P. Krishna, D. Husak, RFID infrastructure, IEEE Applications and Practice, 1 (2) (2007) 4–10.

[18] B. Can, M. Portalski, H.S. Denis, S. Frattasi, H.A. Suraweera, Implementation issues for OFDM-based multihop cellular networks, IEEE Communications Magazine, 45 (9) (2007) 74–81.

[19] J. He, K. Yang, K. Guild, H.-H. Chen, Application of IEEE 802.16 mesh networks as the backhaul of multihop cellular networks, IEEE Communications Magazine, 45 (9) (2007) 82–90.

[20] A. Boukerche, H.A.B.F. Oliveira, E.F. Nakamura, and A.A.F. Loureiro, Localization systems for wireless sensor networks, IEEE Wireless Communications, 14 (6) (2007) 6–12.

[21] I.F. Akyildiz, T. Melodia, K.R. Chowdury, Wireless multimedia sensor networks, IEEE Wireless Communications, 14 (6) (2007) 32–39.

[22] J. Cho, Y. Shim, T. Kwon, Y. Choi, S. Pack, S. Kim, SARIF: a novel framework for integrating wireless sensor and RFID networks, IEEE Wireless Communications, 14 (6) (2007) 50–56.

[23] S. Faccin, C. Wijting, J. Kneckt, A. Damle, Mesh WLAN networks: concept and system design, IEEE Wireless Communications, 13 (2) (2006) 10–17.

[24] J. Jun, M.L. Sichitiu, The nominal capacity of wireless mesh networks, IEEE Wireless Communications, 10 (5) (2003) 8–14.

[25] I. Chlamtac, M. Conti, J.J.N. Liu, Mobile ad hoc networking: imperatives and challenges, Ad Hoc Network Journal, 1 (1) (2003) 13–64.

[26] S. Chakrabarti, A. Mishra, QoS issues in ad hoc wireless networks, IEEE Communications Magazine, 39 (2) (2001) 142–148.

[27] T.J. Kwon, M. Gerla, Clustering with power control, Proceedings of the IEEE Military Communications Conference, 2 (1999) 1424–1428.

[28] A.B. McDonald, T.F. Znati, A mobility-based framework for adaptive clustering in wireless ad hoc networks, IEEE Journal on Selected Areas in Communications, 17 (8) (1999) 1466–1487.

[29] M.L. Torres, C. Vargas, Evaluating connectivity and quality in ad-hoc networks through clustering and trellis algorithms, Proceedings of the 14th International Conference on Electronics, Communications and Computers, 6 (2004) 132–137.

[30] C. Vargas, A. Lopez, Evaluating connectivity in wireless ad-hoc networks, Proceedings of the International Conference on Communications, 6, June (2004) 3613–3617.

[31] J.A. Stine, G. de Veciana, A paradigm for quality-of-service in wireless ad hoc networks using synchronous signaling and node states, IEEE Journal on Selected Areas in Communications, 22 (7) (2004) 1301–1321.

[32] D. Niculescu, B. Nath, Ad hoc positioning system (APS), IEEE Global Telecommunications Conference, 5, November (2001) 2926–2931.

[33] D. Niculescu, B. Nath, Ad hoc positioning system (APS) using AOA, Proceedings of the IEEE INFOCOM, 3 (2003) 1734–1743.

[34] Y. Shang, W. Ruml, M.P.J. Fromherz, Positioning using local maps, Ad Hoc Networks, 4 (2006) 240–253.

[35] E. Ekici, S. Vural, J. McNair, D. Al-Abri, Secure probabilistic location verification in randomly deployed wireless sensor networks, Ad Hoc Networks, 6 (2008) 195–209.

[36] Y. Zou, K. Chakrabarty, Target localization based on energy considerations in distributed sensor networks, Ad Hoc Networks, 1 (2003) 261–272.

[37] N. Patwari, A.O. Hero, M. Perkins, N.S. Correal, R.J. ODea, Relative location estimation in wireless sensor networks, IEEE Transactions on Signal Processing, 51 (8) (2003) 2137–2148,

[38] R. Stoleru, J.A. Stankovic, S.H. Son, On composability of localization protocols for wireless sensor networks, IEEE Networks, 22 (4) (2008) 21–25.

[39] M. Porretta, P. Nepa, G. Manara, F. Giannetti, Location, location, location, IEEE Vehicular Technology Magazine, 3 (2) (2008) 20–29.

[40] G. Sun, J. Chen, W. Guo, K.J. Ray Liu, Signal processing techniques in network-aided positioning, IEEE Signal Processing Magazine, 22 (4) (2005) 12–23.

[41] F. Gustafsson, F. Gunnarsson, Mobile positioning using wireless networks, IEEE Signal Processing Magazine, 22 (4) (2005) 41–53.

[42] M.Z. Antonio, A theoretical framework for the evalution of connectivity, robustness and reachability in wireless ad hoc networks, M.Sc. Thesis, ITESM-Monterrey, 2005.

Satellite-Based Location Systems

Great technologies and applied sciences have great beginnings; the beginnings of the science of positioning can be found in global navigation satellite systems (GNSSs). In fact, positioning became a very popular technological concept that started with GNSS when for the very first time a true global position fix was put into practice, furthering the deep scientific concepts involved, such as general relativity, that must be applied to achieve the high precision that is pursued. At that time, information and communications technologies were already very advanced in achieving GNSSs to find one's position location (PL).

Despite the great results, an unresolved challenge was still pending: PL indoors. This was due to the lack of an ability to recuperate signals in closed environments where satellite signals are obstructed or impossible to detect properly inside buildings, concrete canyons in cities, dense and large flora, or in water where the water column is deep enough to attenuate electromagnetic waves.

In this chapter, we will review the main concepts behind the GNSS. In Sections 7.1 through 7.3, the concepts behind satellite positioning, such as those used for any satellite system for space positioning, are described. We will provide a comprehensive development of the technological and scientific ideas and also give the basis for the possible creation of a proper and novel business in this field. Section 7.4 presents modern and general applications of satellite positioning. Sections 7.5 and 7.6 provide a description of how the expected high-accuracy positioning is degraded, allowing the reader to understand various practical implementations around the world. Also in this chapter, the basics of augmented systems are presented. Finally, Section 7.6 projects future developments.

7.1 SATELLITE POSITIONING

Electromagnetic waves used for communicating human information should travel in the so-called line of sight (LOS)—that is, the shortest distance in space-time geometry where transmitter and receptor are embedded. With this said, we must stress the idea of not just considering straight lines but rather all natural modern environments that modify the geometry of the space-time domain. When electromagnetic waves were placed in satellite environments, the ability to detect them easily around the globe became a very powerful tool in communicating information; their use in navigation at sea was one of the most practical applications. Therefore, position fix pertains not only to people but to objects as well. Let us call each fix a *happening* because in fact it may occur at a specific point in space and time regardless of the object of interest. A happening includes stationary or moving objects that are clearly identified as a point or a set of points that form an "extended" happening.

Satellite positioning basically refers to determining the position $\mathbf{R}(t)$ of a happening when the position $\mathbf{r}(t)$ of a satellite is known. Figure 7.1 shows the relationship among $\mathbf{R}(t)$, $\mathbf{r}(t)$, and a measured range vector $\mathbf{e}\rho(t)$ between the two.

If three coordinates are sought, then at least three satellites ought to be in sight. If time is to be determined as well, then four satellites are a must. In each case electromagnetic waves are used for ranging the necessary satellites. When visible light is used, ranges are called laser ranging; when frequencies lie in the radio band, it is radio ranging. Radio ranging may occur at the mobile band (L band from 1.2 to 1.6 GHz) or at VHF and UHF bands (from 150 to 400 MHz) for practical Doppler measurements. Radar ranging is also available when C and S bands are used. The S band also provides the option of using very long base interferometry (VLBI), which results in the most accurate position fix. Interferometry measurements are mainly related to the observable group

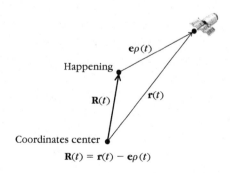

$$\mathbf{R}(t) = \mathbf{r}(t) - \mathbf{e}\rho(t)$$

FIGURE 7.1

Positioning a happening.

delay of a front wave and the delay rate. In any instance where radio detection is used, correction of the delay due to the ionosphere is done through two radio frequencies. In the VLBI case, one frequency is in the S band (2.2 to 2.3 GHz) and the other is in the X band (8.2 to 8.6 GHz).

Range measurements are done indirectly through observable quantities that are named the pseudoranges. The most used pseudoranges are t, the time it takes an electromagnetic wave to travel from a satellite to a receiver, and the *phase observable*, which is the phase of the wave.

Laser ranging is not very practical because it depends on the level of transparency of the atmosphere for visible light. In spite of the enormous possibilities of radio links, they still present many difficulties.

We will discuss how select problems were overcome and how new challenges may appear due to changing frames of reference.

7.1.1 Absolute and Relative Positioning

Absolute positioning occurs when there is a common frame of reference and the happening is referenced to such a frame. Here, all surrounding happenings are assumed unchanged in space and time, therefore, absolute positioning is only valid for one specific interval (in the space-time sense). Strictly speaking, it is also a relative positioning because the frame of reference is not absolute (see Figure 7.2).

We have used absolute positioning here instead of point positioning because happenings have always been described in the space-time domain. Relative positioning refers to fixing a happening with respect to another happening. In the context of considering many changes in the environment, this location position may become difficult. In a simple model in which calculations are made, some of the changes may not be considered since the happenings share the same natural phenomena, as shown in Figure 7.3.

Thus, if we know the position $\mathbf{r}_1(t)$ of a given happening, then the inter-happening position vector $\Delta\mathbf{r}_{12}(t)$ is determined by the equation $\Delta\mathbf{r}_{12}(t) = \mathbf{r}_2(t) - \mathbf{r}_1(t)$ where $\mathbf{r}_2(t)$ is the position of the second happening in the same frame of reference.

In the case of $\mathbf{r}_2^L(t)$ being determined in a different frame of reference—let us say the local reference system and thus the superindex L—then $\mathbf{r}_2(t) = \mathbf{C} + \xi\mathbf{R}(r, p, y)\mathbf{r}_2^L(t)$ where $\mathbf{C}$ is the translation vector describing the origin of the local frame of reference with respect to the original frame of reference, ξ is a scale factor, and $\mathbf{R}(r, p, y)$ is the rotation matrix given by the misalignment angles between the two Cartesian frames and characterized by rotations known as roll, pitch, and yaw. Note that for successive transformations, the rotation matrix results from multiplying all matrixes involved in pairs of frames of reference.

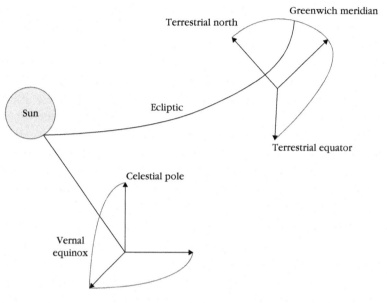

FIGURE 7.2

Absolute positioning.

The very first inertial frame of reference was the celestial one, followed by the terrestrial frame, then the local frame of reference, and finally the object frame of reference. In geomatics or geospatial sciences, special reference frames called *datums* are surfaces of constant coordinate values. The most popular in classical geography is the geoid, vertical datum, given by the equipotential surface of gravity field, basically mean sea level. If we best fit a geometrical Cartesian surface to geoidal data, then an ellipsoid is what we obtain, a horizontal datum, which represents the popular and widespread latitude λ and longitude φ coordinates (see Equation 7.3).

There are several horizontal datums because they depend on the origin of the Cartesian coordinate system. Although it is quite difficult to define a unique center of coordinates (whether geocentric or nongeocentric), there are universally accepted horizontal datums. Examples of evolutionary horizontal datums are WGS84 and ITRF 90, and they are related by the Helmert transformation, as described in McCarthy [5]:

$$
\begin{bmatrix} x \\ y \\ z \end{bmatrix}_{ITRF90} = \begin{bmatrix} 0.060 \\ -0.517 \\ -0.223 \end{bmatrix} + 0.999999989 \begin{bmatrix} 1 & -0.0070'' & -0.0003'' \\ 0.0070'' & 1 & -0.0183'' \\ 0.0003'' & 0.0183'' & 1 \end{bmatrix} \begin{bmatrix} x \\ y \\ z \end{bmatrix}_{WGS84}
$$

when small roll, pitch, and yaw occur.

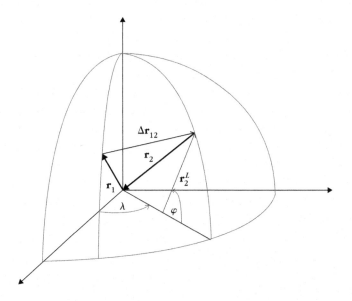

FIGURE 7.3

Relative positioning.

To determine the PL of a happening, when and how must be decided. The reference system for time still remains to be defined. As mentioned before, the very first global reference system was the Conventional Celestial Reference System (CCRS) followed by the Conventional Terrestrial Reference System (CTRS), which is attached to the Earth.

The CTRS has variations because of pole movements, which are in turn due to changes in the Earth's crust (tectonic plate motion and earth tides) and other terrestrial deformations such as those made by ocean loads. Similarly, the Celestial Ephemeris Pole moves with respect to the North Ecliptic Pole. Obviously those pole movements are detected with respect to CCRS. Thus, all known and quantifiable variations must be considered if we want to be precise.

The global reference for time is the Universal Time Coordinate (UTC) and it is fixed to the CCRS. Meanwhile the time fixed to the CTRS is the Greenwich Apparent Sidereal Time, which is referenced to the orbital positions of the Earth around the Sun, or ephemeris time (ET). Initially, this time was called terrestrial time (TT) and was measured by an international atomic time (TAI). Later, better measurements and instruments allowed us to define on January 1, 1977, a single Universal Time that is related to TAI by the equation $ET - TAI = 32.184$.

When UTC is applied, in general it needs to be adjusted in leap seconds, while ET does not. Furthermore, corrections should be kept in mind for differences among all possible calendars or origins of reference time, such as between Julian and Gregorian. Other considerations for corrections are related to different models for nutations and precession movements.

Alongside the concept of relative positioning we find a very similar concept of differential positioning. Both make use of two happenings to identify the coordinates for one of them. Differential positioning employs pseudorange corrections, while relative positioning calculates the relative positions between the two happenings, that is, its base line.

Specifically, relative positioning uses a time of reference t_0 for observations in both happenings at which the phase carrier is received. Calculations can be made while the observation time span $t - t_0$ is coherent, which is also known as latency time (latency, in fact, gives the size of the surroundings for the time frame of reference). Relative positioning is an order of magnitude higher than differential positioning; furthermore, differential positioning can be obtained from relative positioning when latency is zero.

Differential positioning uses either code-phase pseudoranges or carrier-phase pseudoranges, and when a quantity called *ambiguity* [8] is resolved, the precise differential is obtained. Table 7.1 shows a comparison of precision between conventional and precise differential positioning.

Differential positioning may also use one or more reference stations, forming single or multiple differential positioning. Relative and differential positioning need a way to link the happenings involved. Thus, several radio links, architectures, and communications formats have been developed. Among them we can find cellular networks, amateur radio links, and even radio broadcasting of terrestrial, maritime, or Internet links that transmit radio packets under standard formats. A more recent GNSS data transmission makes use of terrestrial pseudosatellites, known as *pseudolites*. For high-precision PLs, data transmission rates and synchronization are important.

7.1.2 Kinematic and Static Positioning

Related to time only and minimizing the effects of space, PL in an unchanged time is what is called *static positioning*. Model equations and solutions for this

Table 7.1 Differential Positioning, Conventional and Precise

Characteristics or Technique	Conventional	Precise
Observable	Code pseudorange	Phase pseudorange
Application range	100 km	10 km
Horizontal precision	less than 1 m	less than 10 cm

case may be written in general form and can be easily calculated using high school geometry equations.

When considering movement of the happening, either in absolute or relative positioning calculations, care must be taken. Further, the happening of interest and its reference can be changing in a more general scenario. Existing applications are still far removed from relativistic considerations, but anyone interested in visionary business or actual military applications are advised to use more advanced theoretical effects in their mathematics and models.

In position fixing, we need quantities called *observables*. Among the observables are direction, ranges, pseudoranges, range differences, and range rates. When we take into account range rates, the process is called kinematic positioning. Depending on the PL procedure, some of these observables are used and some are discarded.

One of the main observables is the range given by

$$\rho(t) = \|\mathbf{r}(t) - \mathbf{R}(t)\|, \tag{7.1}$$

where $\|\|$ is the norm of the vector $\mathbf{e}\rho(t)$. Pseudorange is an estimate of such quantity, but is affected by a bias $\Delta\rho$ produced by the lack of synchronization between the time of emission of the signal and the time of the happening. Therefore, its mathematical expression is

$$\rho(t) = \hat{\rho}(t) + \Delta\rho. \tag{7.2}$$

In 3D range determination, we need four pseudoranges to determine the three coordinates and the bias. Note that the bias is in fact related to the time coordinate because that is the real quantity measured. Pseudorange can be interpreted as a spherical shell of radio $\hat{\rho}(t_0)$ centered in the happening at instant t_0, and the unknown position is in a sphere of radius $|\Delta\rho(t_0)|$ delimited by the pseudorange shell. When considering a constant difference of pseudoranges between two reference points and a new location of the happening (a new point in the happening's world line), they are in turn considered to be in a hyperboloid that has as its foci the two reference points. A 2D representation of the corresponding hyperbolic curves appears in Figure 7.4.

When using four satellites as four reference points, which are in a circular cone with respect to the reference system, the unknown location is at its apex. Then the system of equations that describes the system becomes difficult to solve because the solution matrix is ill conditioned. The corresponding geometry can be evaluated as an index called dilution of precision (DOP). When the value of DOP is small, the geometry is improved; when DOP tends to infinity, then the geometry approaches the cone configuration. Along a world line, various DOP factors can be associated with a new index, which is referred to as position dilution of precision (PDOP).

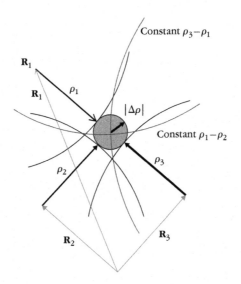

FIGURE 7.4

Pseudoranges and uncertainties.

When a stronger geometric configuration is pursued, it is common for surveyors to link several control points, which are point positioned with high precision. Thus, a network is formed and positions are transmitted across the network. On one hand, random errors are averaged since the ecosystem is mixing its variables; on the other, systematic errors tend to propagate. A surveying campaign normally makes a plan of how to cancel or minimize errors [1]. When horizontal positioning is more important than knowing the height, the network is a horizontal geodetic network; the opposite situation is the geodetic labeling network.

Kinematic positioning refers to calculating the velocity of the happening, which can be obtained by differentiating Equation 7.1, as in

$$\dot{\rho}(t) = \|\mathbf{r}(t) \dot{-} \mathbf{R}(t)\|,$$

or in vectorial form,

$$\dot{\rho}(t) = \mathbf{e} \cdot \|\mathbf{r}(t) \dot{-} \mathbf{R}(t)\|. \tag{7.3}$$

Since **e** is the unit vector that joins the happening with the reference point, Equation 7.3 represents radial velocity. Thus, radial velocity is the projection of $\|\mathbf{r}(t) \dot{-} \mathbf{R}(t)\|$ onto the vector $\|\mathbf{r}(t) - \mathbf{R}(t)\|$. Geometrically, it represents a circular cone with its apex at the reference point and its axis along $\|\mathbf{r}(t) \dot{-} \mathbf{R}(t)\|$. In practice, radial velocities can be easily determined by Doppler measurements of the radio wave that links the happening with a satellite on the GNSS.

Normally, the happening is of a space-like type, which implies that the trajectories are spatial curves on a curved surface. In this case, we have two possible curves: geodesic and loxodromic. When terrestrial trajectories are of interest, then the curved surface can be approximated by a sphere rather than an ellipsoid. In a sphere, loxodromic becomes a constant geodetic direction; meanwhile, geodesic is named orthodromic. In maritime trajectories, the orthodrome is better described by sequential loxodromes.

For a better understanding of happening trajectories, we need to use the differential geometry of space–time.

7.2 STRUCTURE OF A SATELLITE-POSITIONING SYSTEM

The boom in satellite positioning started at the end of the 20th century with the global positioning system (GPS). Since then, many other GNSS have been planned or developed. All existing and modified systems that may be conformed into a single system [2] use the same basic concepts and therefore the same structure. Figure 7.5 shows what is needed in any satellite system.

Examples of GNSS are the Russian GLONASS and the European system GALILEO. Each group is in charge of its own system and is responsible for

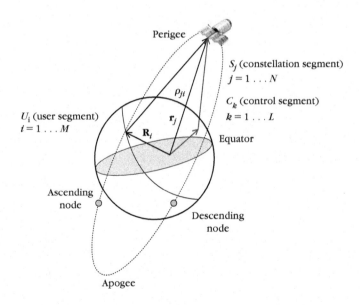

FIGURE 7.5

Elements of a satellite system.

maintaining, updating, and improving it, and also for informing all interested parties of any changes in system status.

Communication to (uplink) and from (downlink) the satellite may occur using different techniques. These techniques and their general principles are mentioned in the following description of satellite signal structure.

7.2.1 Constellation Segment

For PL purposes, to cover in a predefined manner specific areas of space-time, several satellites are placed in Earth orbit. The specific deployment of satellites conforms to the constellation segment. Regarding coverage needs, satellites are placed in low Earth orbit (LEO), medium Earth orbit (MEO), or high Earth orbit (HEO), and all of them obey Kepler's Third Law.

The following equation for a satellite orbiting Earth can be obtained from celestial mechanics [3]:

$$\ddot{\mathbf{r}} = -G(M + m)\frac{\mathbf{r}}{r^3},\tag{7.4}$$

where G is the gravitational constant, M is the Earth's mass, m is the satellite's mass, and $\mathbf{r}$ is the vector that joins both objects. Finding the solution to the differential Equation 7.4 will give a family of orbits with different shapes, sizes, and orientations. In terms of $\mu = G(M + m)$ and from the geometry of the system, the general solution in Equation 7.4 represents a conic section, and when described in polar coordinates takes the form

$$r = \frac{k^2/\mu}{1 + e\cos\delta},\tag{7.5}$$

where k is the angular momentum measured in a geocentric frame divided by the planet's mass, e is a vector defined by $e = -\frac{\mathbf{r}}{r} - \frac{1}{\mu}(k \times \dot{\mathbf{r}})$, and ζ is the true anomaly defined by $\mathbf{r} \cdot e = ||r|| \, ||e|| \cos\zeta$.

A full physical and geometric interpretation comes from the fact that $k \cdot e = 0$ in agreement with Figure 7.6, and the total energy of the satellite

$$mh = \frac{1}{2}mv^2 - \frac{m\mu}{r}$$

is related by $\mu^2\left(e^2 - 1\right) = 2hk^2$.

Based on all of these equations, the meaning of e in Equation 7.5 is that of eccentricity of the conic, which is the magnitude of the vector e. Thus, when $e = 0$ we have a circle, $0 < e < 1$, which is an ellipse, and the other two solutions, which are less convenient for a satellite—$e > 1$, a hyperbola, and $e = 1$, a parabola.

As is well known and expected, the most common trajectory for a satellite is that of an ellipse, which coincides with Kepler's First Law. This law, adapted to satellites, reads: "The orbit of a satellite is an ellipse, one focus of which is the

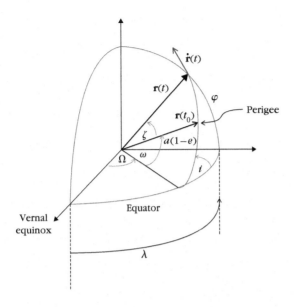

FIGURE 7.6

Orbital elements.

Earth." For further understanding of the physics and geometry, $p = k^2/\mu$ must be defined as an orbit parameter in Equation 7.5, which at the same time can be given in terms of the semimajor axis as $a = \frac{p}{|1-e^2|}$. Also, from the energy relationship given above, the size of the orbit is related to the energy. An ellipse is defined when $a = -\mu/2h$, and a hyperbola is defined when $a = \mu/2h$. Thus, when the energy is positive, the satellite has enough energy that it recedes without boundary, and if $h = 0$, we have a parabola. Neither a parabola nor a circle exists in real life.

Figure 7.6 is referenced to a specific system of reference attached to Earth and therefore Ω, the longitude of the ascending node measured counterclockwise from the vernal equinox, is where the satellite crosses a fixed reference plane. Inclination i gives the obliquity of the satellite orbital plane, and Ω and i determine the orientation of the satellite movement plane, which coincides with the direction of the vector k. Meanwhile, the direction of the perihelion, ω, called the *argument of the perihelion*, gives the same direction of the vector e, which is measured from the ascending node in the direction of motion.

Satellite motion is normally counterclockwise, while the opposite movement results in a retrograde orbit. Both can be used for any GNSS, depending on the programmed coverage. Standard orbital elements ω, i, and Ω constitute the data of the ephemeris of a satellite. *Ephemeris*, in general, is the radio vector of the satellite as a function of time for a specific frame of reference. When a satellite ephemeris is acquired by direct optical measurements, using two angles to define the direction of $\mathbf{r}(t)$ is common: right ascension (α) and declination (δ), which are very well known when using the astronomical right ascension system.

An interesting datum is surface velocity, which is the area swept by the satellite radial vector per unit of time, which in terms of the distance r and the true anomaly ζ is $\dot{A} = \frac{1}{2}r^2\dot{\zeta}$. On the other hand, one can calculate in polar coordinates the magnitude of the vector k as $k = r^2\dot{\eta}$, therefore resulting in $\dot{A} = \frac{1}{2}k$. This, in fact, is Kepler's Second Law for satellites: The radio vector of a satellite sweeps equal areas in equal amounts of time. Kepler's Second Law directly leads to $dA = \frac{1}{2}kdt$, and so when integrating the entire satellite period P and over the total area of the ellipse, it can easily obtain Kepler's Third Law: $P^2 = \frac{4\pi^2}{G(M+m)}a^3$. That is, the ratio of the cubes of the semimajor axes of the orbits of two satellites is equal to the ratio of the squares of their orbital periods. If we use the mean motion $\nu = 2\pi/P$ and the approximation to the reduced mass $\mu \approx GM$, and because m is negligible with respect to M, the usual form for Kepler's Third Law is now $\nu = \left(\mu/a^3\right)^{1/2}$.

From such results, it can easily be seen that depending on the orbit, the time for each pass can be calculated, and from there an orbit can be selected to accomplish the desired coverage. Also, for various GNSS there are different passing times, and thus users can conjugate all those GNSS in a single system. Such combinations will provide better precision and synchronization [4]. However, previous agreements among those systems should be prepared. This is another business opportunity for a task force to locate all technical and human resources for the construction of an integrated GNSS.

In Section 7.3.2, various metrics will be discussed to evaluate performance of a single GNSS or a group of GNSS as a whole system. Kepler's laws govern the movements of all the satellites in the GNSS constellation as shown in Figure 7.7.

Lately auxiliary constellations are being put in place for the improvement of PL. These are called *augmented systems*, and they are both terrestrial based and satellite based. Knowledge of satellite constellation is of vital importance because all calculations are done from them. The position of satellites in a 3D graph is what is known as a constellation ephemeris, which is referenced from a specific terrestrial frame of reference. It is also useful to have a 2D projection on a flat map called *satellite ground tracks*. Ground tracks are shown in Figure 7.8. Other satellite plots show satellite positions, called *satellite polar plots*, and are used, for example, by GPS receivers as is depicted in Figure 7.9.

For high-precision calculations, Kepler's equations should be altered by the inclusion of many other forces acting upon a satellite. Included here are gravitational forces related to the Sun, moon and planets; solar radiation pressure and magnetic forces; and even forces that are so small, one can think they have no

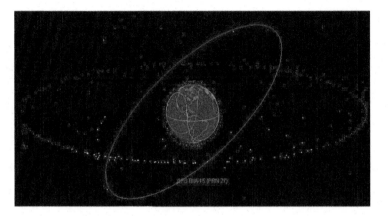

FIGURE 7.7

Constellation (see *http://science.nasa.gov/realtime* for a color version of this image and Figure 7.8).

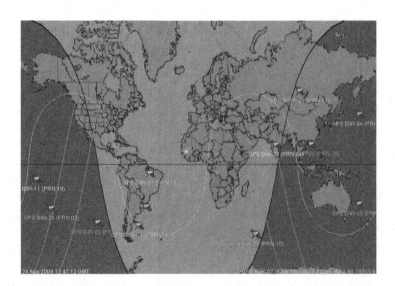

FIGURE 7.8

Ephemeris in 2D (*see http://science.nasa.gov/realtime/*).

influence on satellites (e.g., the effects of tidal forces, the Earth's surface deformation, and atmospheric dragging). Although the latter forces are small, their dragging defines the lifetime of a satellite because they are constantly slowing down its movement. Perturbation methods and numerical methods are used to solve the equations; therefore, precision for the ephemeris varies.

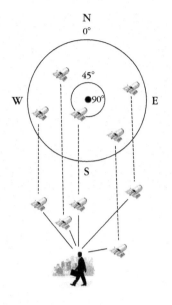

FIGURE 7.9

Polar plots shown in a GPS screen receiver.

An important part of the constellation is the way that satellites communicate with Earth. There are different ways of accessing information coming from satellites that also permit satellite identification. For instance, if satellites transmit information at different times, time division multiple access (TDMA) is the method used; if the frequency of the satellite radio link is used instead of time, the technology is called frequency division multiple access (FDMA); if there is a code impressed in the radio link for each satellite, the method is code division multiple access (CDMA). In the latter case, CDMA, the original signal spectrum density suffers a spreading in the frequency domain. Such an effect produces so-called spread-spectrum signaling, and it is achieved by the use of a pseudorandom noise code. Pseudorandom noise indicates that the code has very specific statistical properties; on one hand it has almost no information, and on the other, the autocorrelation is almost zero except at zero delay. This spread spectrum is done in a very simple form by binary biphase modulation of a carrier, known as BPSK. Radiating two radio signals in different bands with the purpose of compensating the effect of the ionosphere is common.

The use of CDMA also allows transmission of various qualities of information just by changing the pseudorandom codes. Different names are given to different codes; for example, in GPS there are two codes: CA and P, which is sometimes called Y. All are oriented to different users and at the same time they are basically acquired at different speeds for security reasons as well.

7.2.2 Control Segment

In any communications system, control must exist. In the case of a satellite system the control serves as a synchronized system as well. By synchronized, we mean that the time part of the space-time geometry may happen under a specific precision. Most of the GNSS use time as an important part of their PL; therefore, there are two main purposes of the control segment: (1) to keep a very precise atomic clock and to synchronize the entire system, and (2) to keep track of the position of each and every satellite of the constellation.

Tracking of PL is very challenging, and ways to improve such tracking with fast and easy tools is an interesting area of study. One can argue that given Kepler's laws, all necessary ephemeris data are already calculated. However, it is quite impossible to get true positions in real life for each satellite because motion equations should include the whole system of satellites with its variations in distances, speeds, and relativistic corrections, which are due to variations in the gravitational field. These variations are so numerous that there are several methods per mathematical model from numerical to analytical.

Generally, several tracking stations should be spread around Earth such that satellite control and the gathering of information from the constellation can be done at all times. Care must be taken when calculating satellite orbits because there are at least two different frames of reference: the RA system, which states the position $\mathbf{r}$ of the satellite, and the conventional terrestrial system, which is the one used by the tracking stations with vector position $\mathbf{R}(\varphi, \lambda, h)$. We can calculate the topocentric range as

$$\rho = \|\mathbf{r} - \mathbf{R}\|, \tag{7.6}$$

where $\mathbf{r} = \mathbf{R}(\theta)\mathbf{R}_{xx'}\mathbf{r}'$ is being calculated through the rotations in order to obtain the RA system from the CT system (Figure 7.10).

The rotation matrix $\mathbf{R}(\theta)$ is achieved through the Greenwich apparent sidereal time θ. In this case, the polar motion has been ignored. It is easy to see that the rotation rate of the Earth is $\frac{\partial \theta}{\partial t} = w_e \approx 7.29211515 \cdot 10^{-5}\,\mathrm{rad/seg}$ in the Earth ellipsoid. Also, if we define the radius of curvature of the Earth ellipsoid as $\varrho = \frac{a}{\sqrt{1-e^2 \sin^2 \varphi}}$, then

$$\mathbf{R} = \begin{bmatrix} (\varrho + h) \cos \varphi \cos \lambda \\ (\varrho + h) \cos \sin \lambda \\ \left[\varrho + \left(1 - e^2\right) h\right] \sin \varphi \end{bmatrix}. \tag{7.7}$$

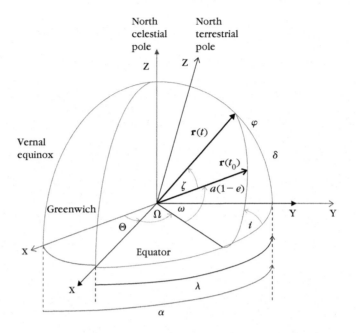

FIGURE 7.10

Celestial and terrestrial coordinate systems.

Thus, the range rate is

$$\frac{\partial \rho}{\partial t} = \left[\frac{\partial \mathbf{r}}{\partial t}\right]^{T} \frac{(\mathbf{r} - \mathbf{R})}{\rho},\tag{7.8}$$

where

$$\frac{\partial \mathbf{r}}{\partial t} = \frac{\partial \mathbf{R}(\theta)}{\partial \theta} \mathbf{R}_{xx'} \mathbf{r}' \frac{\partial \theta}{\partial t} + \mathbf{R}(\theta) \mathbf{R}_{xx'} \frac{\partial \mathbf{r}'}{\partial t} \quad \text{and} \quad \frac{\partial \mathbf{r}'}{\partial t} = \frac{va}{1 - e \cos \delta} \begin{bmatrix} -\sin \delta \\ \sqrt{1-e^2} \cos \delta \\ 0 \end{bmatrix}.$$

Some of the GNSS even have active control systems, which comprise a network of control stations to continuously track each satellite on the GNSS.

The control segment generally uploads the best estimate ephemeris to the satellite, such that the satellite segment can broadcast the data to the user segment. Such ephemeris may contain e, eccentricity; a, semimajor axis; i_0, inclination at ephemeris reference time t_{oe}; $\dot{i}$, rate of inclination; ω, argument of perigee; Ω_0, right ascension parameter; $\dot{\Omega}$, rate of right ascension; Δn, mean motion difference; M_0, mean anomaly; and other data to make corrections in latitude, orbital radius, and inclinations.

7.2.3 User Segment

The other end in a GNSS is precisely the one that uses the PL information. All the points that use this information form the user segment. Basically, this is the set of all happenings that have a GNSS satellite receiver. This segment has a large variety of possible applications, and thus it is where many companies are directing their efforts for innovations. This part is the only one that is not the responsibility of the system's owner; improvements and developments have been left to independent entrepreneurs. The user segment is basically formed by large numbers of receivers with all sorts of characteristics and precision, and therefore prices. Sometimes organizations or nations set up a system of several receivers to form a network and improve precision by relative positioning.

The receivers have many ways to implement radio signal detection. Any GNSS receiver contains antennas; electronic hardware; receiver hardware itself; RF components; communication channels; techniques to spread and despread radio signals; a way to deal with synchronization, stability, acquisition, and access to satellites either continuously or by switching; single or dual frequency detection; and PL techniques through pseudorange or carrier beat. But more important, they are programmed with ways to solve the observation equations.

7.3 FUNDAMENTAL CONCEPTS

In this context, space–time is the actual fundamental concept and the theory of relativity is an inner concept that allows the definition of a happening. From a more traditional point of view, locating a happening in space is the goal behind determining a PL of one or more points in the same space interval. Thus, the calculated distance is what is understood as ranging. Yet, from another view, finding relative displacement in space is directly linked to a relative time. Such time is the time when the signal leaves its origin and the TOA at some receiver, and it is calculated by a master clock, which in the end synchronizes all calculations. These two concepts, ranging and timing, along with how they are inevitably related due to the indivisibility of their nature, are described below.

Another important concept in wireless radio communication is the use of electromagnetic waves. Its mathematical representation is the function $s(\mathbf{r}, t)$, which depends on space and time in the form of

$$s(\mathbf{r}, t) = A \cos (\omega t - \mathbf{k} \cdot \mathbf{r} + \phi_0), \tag{7.9}$$

where A is the wave amplitude, ω is the radian frequency, $\mathbf{k}$ is the propagation vector, $\mathbf{r}$ is the radio vector from the satellite to the receiver, t is the elapsed time to travel the distance r, and ϕ_0 is the initial phase.

7.3.1 Ranging and Timing

When one calculates distance in the purest space-like sense, a comparison exists between the space interval of interest and the reference. In modern times, such reference is a wavelength of well-known oscillatory phenomena. Obviously, in a communications system that uses a fixed oscillatory electromagnetic wave as the radio signal source in which information is transported, the wavelength of such oscillation could be used to calculate the distance. Thus, knowing the number of oscillations that it would take for a signal to travel from a given source to a given receiver will allow one to know the distance within a fraction of the signal wavelength. This method is known as phase ranging because the wavelength of a stationary oscillation is the inverse of its frequency, which at the same time is the rate of change of its phase.

Therefore, from data acquired directly from the arriving wave (see Equation 7.9), one can calculate the distance using the relationship $r = vt$, where v is the velocity of propagation and t the elapsed time; this is known as pseudorange. Pseudorange allows us to calculate the distance in a biased form since time is measured in different time frames, one time frame for the satellite and one for the user. For these calculations, we need accurate information regarding all times involved.

The other technique for indirectly calculating the distance is by making use of $r = \phi/k$, where k is the magnitude of the propagation vector related to the wavelength λ as $k = 2\pi/\lambda$, and ϕ is the wave phase at the instant t. From this, to use the phase and to measure the range, we need to accurately know the satellite frequency and track the phase of the signal continuously.

In the case of coded signals, pseudoranging can be done by using the wavelength of the code, known as a *pulsed ranging*. When the carrier beat phase is used, then we are talking about *continuous wave ranging*. Carrier beat phase results from mixing the incoming Doppler phase-shifted signal with a locally generated signal that has a highly precise frequency.

Two-way and one-way ranging are possible in pseudoranging or phase beat. The capability of doing it in real time or in differing time depends on the electronics used. As was seen in Chapter 2, different techniques can be used for range calculation: TOA, RSS, or AOA. The last two are difficult to use since the received power is extremely low, making it almost impossible to use. Nevertheless, as electronic technology evolves and transits from micro to nano scales, new possible areas in satellite ranging could open.

Meanwhile, TOA is used as follows: the pseudorange t is the difference between the time of transmission measured as proper time T_e and the TOA as a proper time T_r. Those two proper times ought to be referenced to a unique time or reference time τ; therefore, t can be better expressed as a difference, $\Delta t = T_r(\tau_a) - T_t(\tau_e)$, but clearly we can identify a difference between the satellite and the reference time as being $dT_t = \tau_e - T_t(\tau_e)$, and similarly a difference between the receiver and the reference time as $dT_r = \tau_a - T_r(\tau_a)$. Thus, we may have Δt composed by a true difference time $\Delta \tau$ measured from the unique time and a difference $dT_t - dT_r$, resulting in

$$\Delta t = \Delta \tau + dT_t - dT_r. \tag{7.10}$$

The range $\hat{\rho}$ can then be calculated by multiplying that time by the speed of light c:

$$\hat{\rho}_o = c\Delta t = c\Delta \tau + c\,(dT_t - dT_r). \tag{7.11}$$

From there we can identify the true range, $\rho = c\Delta \tau$, which, in fact, is also the true distance given by Equation 7.6 ($\rho = \|\mathbf{r} - \mathbf{R}\|$). Thus, the pseudorange is altered by some statistical bias as a result of $c\,(dT_t - dT_r)$ in Equation 7.11 as well as other sources in the ecosystem, particularly those in the ionosphere b_i and the troposphere b_t. In the end, we have the pseudorange

$$\hat{\rho} = \|\mathbf{r} - \mathbf{R}\| + b_T + b_i + b_t + d\rho, \tag{7.12}$$

where $b_T = c(dT_t - dT_r)$ and $d\rho$ is the uncertainty when measuring $\mathbf{r}$. From Equation 7.11, we can see that synchronization is a must.

Similarly, for the carrier beat phase, we may calculate the observable carrier beat phase. We can write $\phi = \phi_e(T_e) - \phi_a(T_r)$, but $\phi_a(T_r) = \phi_e(T_e) + f\Delta t(T_e - T_r)$, where f is the radio link nominal frequency. Therefore, $\hat{\phi}_o = -f\Delta t$, when using Equation 7.10, becomes

$$\hat{\phi}_o = -f\,(\Delta \tau + dT_t - dT_r).$$

Applying Equation 7.11, we get

$$\hat{\phi}_o = -\frac{f}{c}\hat{\rho}_o.$$

Finally, if we introduce the bias due to the ionosphere, which has an opposite effect like that of the troposphere, we have

$$\hat{\phi} = -\frac{f}{c}\|\mathbf{r} - \mathbf{R}\| - \frac{f}{c}(b_T + b_i - b_t). \tag{7.13}$$

However, a big difference can be found, since phase tracking is not always feasible and in real life there are some cycle slips. Assuming such cycle slips as being an integer of cycles N, or cycle ambiguity, Equation 7.13 gives the total carrier beat phase in the form of

$$\hat{\phi} = -\frac{f}{c}\|\mathbf{r} - \mathbf{R}\| - f(dT_t - dT_r) + N.$$

By defining $\Phi = -\lambda\phi$, we can rewrite the previous equation in units of length as

$$\Phi = \|\mathbf{r} - \mathbf{R}\| + b_T - b_i + b_t + \lambda N. \qquad (7.14)$$

Equation 7.14 is practically the same as Equation 7.12, with the difference being that while in the former equation there is uncertainty in $\mathbf{r}$, the latter contains the ambiguity term.

The terms due to the ionosphere and troposphere can be understood as follows: ionospheric change b_i is calculated as $\int (n-1)ds$, where ds, is the arc length of the trajectory and n is the index of refraction. We can identify two velocities in a wave: group velocity and phase velocity. From there we have two indexes of refraction, the group index of refraction and the phase index of refraction. Refraction indexes are given by

$$n \approx 1 \pm \frac{a\eta}{f^2},$$

where a is a constant, η is the electron volumetric density, and f is the signal frequency. We use the $+$ sign for the phase index of refraction and the $-$ sign for the group index of refraction. Therefore, b_i can be written as

$$b_i = \pm \frac{a\sigma}{f^2}, \qquad (7.15)$$

where σ is now the electron surface density along the path of integration.

From Equations 7.1 and 7.3, it is apparent that space–time appears naturally. Thus, if one considers many errors due to uncertainty in time, power detection becomes more adequate in calculating ranges because power implicitly considers time and space together. This can be easily seen because power is an expectancy value:

$$P = \lim_{T \to \infty} \frac{1}{2T} \int_{-T}^{T} \|s(\mathbf{r}, t)\|^2 \, dt.$$

In this way, many sources of error are considered as part of the signal's ecological system.

7.3.2 Precision and Accuracy

The signal carrier gives the most precise PL; for example, given a wavelength of 20 cm for a 1.5-GHz carrier, we may have a precision of that magnitude. However, as a result of many sources of error, signal carrier precision is difficult and thus precision degrades. Although precision can be very high, as Equations 7.12 and 7.14 show, there are still many uncertainties, mainly because receivers lose cycles in detection.

From these equations, one of the terms used to refer to how precision is degraded is the dilution of precision (DOP), which is nothing more than the root-sum-squared measurement of the size of a confidence region where one wants to place a preselected level of probability. In a statistical sense, DOP is the square root of the traces of various submatrixes of the covariance matrix divided by the measured standard deviation, where the confidence regions are related to the eigenvalues of such matrix. In short, DOP relates the calculations with the geometry of the section of space where satellites are.

In practice, we want DOP to be the smallest possible figure, which means that the geometry of the satellites is as large as possible such that the matrixes will not be ill conditioned. However, with movement of satellites, geometry changes and therefore DOP may deteriorate For example, if we want our positioning precision to be 50 m, for a measured accuracy of 10 m, a DOP of 5 m is required. Afterward, if DOP is 1 m, then we have a positioning accuracy of 10 m. From a purely mathematical point of view, equations can be solved analytically. When using computers to solve equations, the numerical calculations may introduce errors and that could cause the matrixes involved to become ill conditioned.

In GNSS, when performing calculations for PL, there are three different ways of dealing with error correction: one uses only raw code pseudoranges; the second uses the carrier phase information to smooth code measurements; finally, the most precise correction is obtained by using only the carrier phase. When error correction is carried out using carrier phase, we are talking of precise DGNSS. When error correction using raw code pseudoranges or part of carrier phase information, we are dealing with the common DGNSS. To improve the precision in the carrier phase, it is necessary to deal with phase carrier ambiguities. Thus, instead of talking about precision, it is better to talk about accuracy in the statistical sense; that is, the end PL is calculated as an expected value of several measurements.

Care must be taken when a commercial receiver claims very high accuracy because the accuracy depends on how they actually define the dispersion of data; sometimes such dispersion is expressed as being inside 4σ(!) of a region of confidence. Also, each reference axis has different dispersions, which is why using the smallest variance and not the global variance is more marketable.

Also, in trying to reduce sources of error, much effort was invested in trying to reduce uncertainty in knowing the signal ecology system [9]. Thus, ionospheric reduction, channel fading, and other effects are important to predict because they form the ecological surroundings of the communications signals.

7.3.3 Civil and Security Considerations

In a very pure way of thinking, the better the PL information that is known with the highest precision available, the better for all: knowledge and social weave. However, PL as such with its corresponding accuracy is only data, and in this age of information, such data are related to social, political, and military objectives, or in any case economic power. Depending on the use of PL data, some of the data can be less precise because some users do not need very high precision; others, however, need very precise PL for surveillance and security reasons.

7.3.4 Coordinate Systems

We can naively think of a unique coordinate system of reference, which goes in the direction of an absolute way of thinking. Despite all the advantages of a unique coordinate system, there is an advantage to using several references: diverse coordinate systems represent the large diversity of ecological systems for PL and thus allow many fruitful ways of thinking. As suggested in Section 7.1.1, the very existence of a unifying coordinate system is always possible and desirable. But from a practical point of view, a unique ecologic system does not allow us to conceive of different local ecologic systems. Geographic topology, the quantity of electromagnetic signals on communications subspaces, and human constructions and material are some elements that must be known to properly define any local ecologic system, both in size and boundaries.

Local coordinate systems should be accepted and promoted either to calculate the position more precisely or to synchronize happenings. A highly precise PL, known as a local coordinate system, has to be modified to a new concept where any local coordinate system ought to consider the specific communication ecologic system. An interesting and useful task is to develop the mathematical transformations for such new local coordinate systems. Such transformations are not easy due to lack of real models.

Selecting adequate coordinates depends on how many points are collected in the world line and the way in which the collection is done. Barycentric coordinates must be considered either for a compact support set of points or for a discrete set whose density has to be taken into account as well.

7.4 SOURCES OF ERRORS

As was stated, many so-called errors are associated with the impossibility of knowing the source of such errors, and this constitutes a paradigm [6]. Breaking such a paradigm will allow the possibility of innovation. For example, when talking about a happening we have not said anything related to the boundaries of the set of points in the space–time that form such a happening nor have we addressed PL as a result of the solution given for the differential equation of the system. If we define that solution as $\mathbf{r}(t)$, then for such a solution nothing is known about its scale and structure as being a position vector in a metric space or in a subspace of same.

7.4.1 Stochastic

These are true errors, because they have uncertainties that are related to gathering and processing data, and even to the conclusions obtained from the analysis of such data. Stochastic errors have complicated characteristics; however, research may lead to models that permit estimates of their statistical properties. Thus, there are always paths toward improvement.

For example, multipath and imaging of satellite signals are very well-known effects, and efforts are under way to minimize them. Techniques like innovative antenna design and modulation processes are being researched around the world to reduce these errors. But residual error remains due to the lack of exact error models.

Some ideas for improvement include the use of differential positioning. This is an interesting approach and efforts are focused on so-called augmentation systems, which provide a way to improve precision using extra information. Augmentation systems tend to cancel stochastic errors by putting radio stations in the same ecologic conditions as those of the happening.

7.4.2 Systematic

Systematic errors, when detected, predicted, or estimated, are easy to eliminate from calculations. Errors of this type are related to the uncertainty of defining the precise origin of coordinates, such as the exact center of the Earth, when in the CT system, or the repetition of measuring the origin when in the local system, and finally the very center of the happening when in the object system.

The largest errors due to perturbations are related to the third-body or N-body effect, which in general falls on the order of 100 m. This is equivalent to 50 times the figures in Table 7.2 in a time span of 4 hours.

To estimate or predict these errors, knowing the environment where communication occurs is necessary. The difficulty lies precisely in knowing all the elements involved in the environment. Much effort has been focused on obtaining a precise idea of the source of these errors.

Table 7.2 Solar Radiation Perturbations on GPS	
Orbital Element	**Perturbation**
a	5
e	5
i	2
Ω	5
$\omega + M$	10

Another source of error is in the previously mentioned atmospheric drag and solar radiation. Solar radiation has two effects:

1. When incidence is direct, it creates pressure on the satellite. This pressure depends on the surface and mass of the satellite and on whether the satellite is exposed to the Sun or is in the Earth's shadow.
2. When incidence is reflected from the surface of the Earth, it creates the albedo effect.

The force upon a satellite, using the Beer–Lambert law, is given by $bpA\varkappa(\lambda)$, where b is a binary value 1 or 0 depending on whether the satellite is illuminated by the Sun, p is the radiation pressure, A is the cross-section area of the satellite, and $\varkappa(\lambda)$ is the reflectivity of the satellite for the impinging electromagnetic wave with wavelength λ. For example, 4 hours of total solar radiation produces an acceleration average $10^{-7} m/s^2$, and the effect on the satellite's arc length, measured in meters per each orbital element, is presented in Table 7.2.

Meanwhile, atmospheric drag is a typical friction force which is proportional to the velocity of the satellite $\mathbf{v}$, the cross-section area A or rather the effective section and the local atmosphere's density. This force is represented by the mathematical expression $\rho A \mathbf{v}^2$.

Other satellite-related errors come from the ephemeris prediction, which is done in two parts: online using a Kalman filter and offline using a least-squares best-fit curve. These errors are already considered in the ephemeris broadcast of the position of every satellite.

Other station-related errors that are not well known are once again related to frames of reference. The antenna phase center, when it is associated with the emitting antenna, is called the apparent source. This center is not constant but rather is dependent on the emission–reception angle.

Other types of errors are related to time: for example, those resulting from satellite clock drifting produced by changes in temperature and other effects of extreme environments. Drifting error can be calculated and therefore already contained in the ephemeris.

Observation-dependent errors are related to the way that signals are generated and detected. How a particular GNSS is conceived depends on how engineers interpret concepts and even on the way they implement the system as a whole.

All errors mentioned here have been modeled, and theoretically they can be removed; however, some are correlated and not as easy to remove. For instance, let us see how the error due to ionospheric delay can be cancelled because that error depends on frequency (Equation 7.15). When two frequencies are used we can easily compare both pseudoranges using either Equation 7.12 or Equation 7.14, arriving at the so-called ambiguity,

$$d_i(f_1) = \left(\hat{\rho}(f_1) - \hat{\rho}(f_2)\right) \left(\frac{f_2^2}{f_2^2 - f_1^2} \right).$$

Ambiguity can be resolved by using appropriate algorithms and averaging processes.

However, when a single frequency is used we can still cancel ionospheric effects if we have a precise model of the atmosphere. Something similar happens when using tropospheric models because that part of the atmosphere varies more drastically with solar activity and humidity. Opportunities for innovation emerge to cancel errors when two or more observations are used and when single, double, or triple differences are performed, or whatever the model equations indicate.

7.5 APPLICATIONS

In all of the applications that follow, it is important to consider a complete technological package, meaning complementary articles such as maps and nearly everything that we need to consider, always obeying data representation standards. Understanding maps, their elements and projections, is also a must for anyone who wishes to establish new business ventures.

Applications define the precision at which we want to work; once precision is fixed, applications can be defined. In either case, well-planned applications should be programmed into whole systems. Static, kinematic, and differential PL can be used in all the following applications.

The reader should keep in mind that civil applications may define low, medium, and high precision, while the military has higher standards, making the civilian standard high its low. To stay well informed, consider perusing specialized journals and magazines, such as the classical and traditional magazine *GPS World* [12]

7.6 TRENDS AND COMPARISONS

So-called terrestrial positioning systems may include old systems initially created as navigational aids. Such systems were based on radio communications technology and particularly radio direction finding, which used a source of radio waves; at the receiver, there were methods of finding the direction from where the radio wave was coming. The source was basically considered as a beacon to a mobile user, and direction finding was the relative bearing between the mobile user and the beacon.

Another positioning system was based on at least three transmitters, becoming a trilateration hyperbolic system. The most popular positioning systems were based on the Loran and Loran C—or its counterpart, the Russian Chayca system. Precision was 5 to 18 m and up to 90 m RMS in a 2D frame of reference. These systems are still in use today.

Other sources of radio waves for positioning include TV, radio broadcasting, and telephone cellular systems. In these systems, precision is lowered to around 400 m. However, compared to GNSS, they have the advantage of simplicity and low cost.

All previously mentioned systems are oriented to outdoor applications of radio propagation models, and error compensation methods are being studied.

Regarding indoor navigation or PL, many radio propagation techniques are still in development. Systems oriented to indoor PL were described in this book and their use is one of the most modern trends in navigation. Outdoor systems can be compared based on precision, but between outdoor and indoor applications, comparisons are not easy because of the numerous techniques and methodologies involved.

7.6.1 GPS, GLONASS, and GALILEO

At present, there are two well-developed GNSSs, the popular U.S. global positioning system (GPS) and the Russian *Global Náya Navigatsionnaya Sputnikovaya Sistema (GLONASS)*. Europe is also trying to implement its own system named GALILEO. All work in a similar way and on a similar scientific basis; the only differences are in the way that they are implemented in practice (number of orbits and satellites, etc). Thus, the satellite architecture is essentially the same because all of them use control, satellite, and user segments. Table 7.3 presents their practical but not essential differences.

Frames of reference are different. The GPS uses WGS 80, while GLONASS uses PZ-90 ECEF, and Galileo uses GTRF. Transferring data from one system to another is a little laborious because of the different frames of reference.

Table 7.3 Satellite Orbital Parameters for GPS, GLONASS, and GALILEO

Satellite Segment	GPS-M	GLONASS-M	GALILEO
Orbit altitude	20,180 km	19,100 km	23,616 km
Orbital planes	6 separated 60°	3 separated 120°	3
Constellation	28 (28 a 30 *real*)	24 (19 *real*)	30
Orbital inclination	55°	64.85° switched 15°	56°
Orbital period	12 sidereal hours 11:58 *solar hours*	8/17 sidereal hours 11:15:44 *solar hours*	14 sidereal hours 14:04 *solar hours*

Table 7.4 Signal Structure for GPS, GLONASS, and GALILEO

Radio Signaling	GPS	GLONASS	GALILEO
Access Type	CDMA in BPSK CA & P & M codes	FDMA and BPSK CA & P codes	altBOC Polarized BPSK
Services	Standard positioning Precision positioning	Standard precision High precision	Open service Commercial service Safety of life Public regulated Search and rescue
Frequency transmission L band (MHz)	$L1$:1575.42 $L2$:1227.6 $L5$:1176.45	$L1$:1602.0 → 1609.3125 $L2$:1246.0 → 1251.6875 $L3$: centered at 1205	$E5a$:1164 $E5b$:1214 $E6$:1260 → 1300 $E2$:1559 $E1$:1591
Navigational message	Keplerian elements	State variables	Keplerian elements
Information rate	50 sps	50 sps	50,250 and 1000 sps

Also, time references are in principle different, and corresponding care must be taken because they differ only by a few milliseconds. The GPS uses the UTC (USNO) as its time reference; GLONASS, the UTC (SU); and GALILEO, the UTC (BIPM).

The main differences can be found in how the navigation information is sent to Earth, and how available it is for common users. Signal structure elements are presented in Table 7.4. In this table it is worthwhile to note that GALILEO is the most advanced system because its design includes experiences gathered from GPS and GLONASS. Fortunately, those experiences permit improvements on the existing old GNSS, and with them the modernization of GPS. Meanwhile, GLONASS has been updated after the dismantling of the Soviet

Union, and the Russian Federation is still working on a fully operational system. The exact status of all GNSS can be monitored in specialized journals such as *GPS World* [10].

Another advantage of the existence of all three systems is that they can work together, forming a true GNSS. Interoperability is important for novel applications. An example of a full GNSS receiver is the TOP CON GR-3.

As a practical rule of thumb, a realistic official precision for each system follows: horizontal (100 m) 2.5 m, vertical (156 m) 4.5 m, and time (334 ns) 5.7 ns. Higher precision can be achieved using improvement techniques described previously.

7.6.2 Developments in Perspective

Old navigation systems such as the Loran C are being used to transmit modern navigation data simultaneously with their normal signals. They do not interfere with each other, but rather contribute navigation information integrity. Modern navigation uses not just the Loran C satellites but also the Inmarsat satellites of the 80-nation consortium, International Mobile Satellite Organization (IMSO).

There are other GNSS that are not as well known because they are oriented to specific objectives: ARGOS is oriented to environmental research; COSPAS-SARSA, mobiles in distress; DORIS, PL with a very high precision of fixed points on the surface of the Earth; GEOSTAR/LOCSTAR, a two-way positioning system; and EUTELTRACS, fleet control and management.

There are other satellite systems that have been developed to supplement existing positioning systems and reduce errors in such a way that precision can reach millimeters. Again, it is basically a differential positioning and the helper satellites' main duty is to broadcast correction messages for a very wide area. These systems belong to various countries, including:

- The U.S. Wide Area Augmentation System (WAAS).
- The European Geostationary Navigation Overlay System, which uses both GPS and GLONASS, and in the near future will use the GALILEO GNSS.
- The Japanese Multifunctional Transport Satellite Satellite-Based Augmentation System (MSAS or QZSS) with its purpose being to serve the Asia–Pacific region.
- The Canadian Wide-Area Augmentation System (CWAAS), used as a complement to WAAS.
- The Satellite Navigation Augmentation System (SNAS or COMPASS) in China.
- The Indian Augmented Navigation System (GAGAN) that works with GPS and GEO.

7.6.3 Integration of Satellite and Ground-Based Location Systems

Integration may occur in two areas: (1) in the way wireless communication signals are detected (TOA, AOA, RSS), and (2) in conjugating both outdoor and indoor systems.

Perhaps the very first ground-based system developed to help satellite positioning was related to differential positioning since two or more ground stations of a network of reference allow more precise PL. Ground stations for this purpose transmit their data corrections through radio broadcasting. Obviously, improving precision calculations is related to the distance between the receiver happening and the reference point of the network.

In the case of a one-point ground reference, there are three different approaches to obtain the correction data:

1. The reference point sends correction data to the happening of interest.
2. The modeled parameters in a state-space reference network are transmitted to the happening.
3. A reference receives raw data from the happening and then corrects it locally.

Differential corrections are broadcast in RTCM format and services may provide corrections for inland, shore, or maritime users at different rates from 50 to 200 bps. The precision attained by these methods ranges from 10 m to 10 cm [4].

Examples of the first type are the augmentation systems that provide a way to improve precision using extra satellites. Meanwhile, one-point reference is easier to work with compared to the network, which is complex and costly.

Another way to improve PL precision is the use of ground-based transmitters that send signals similar to those of GNSS, and together with the true satellite signals calculations are made based on total signals. The advantage of such ground transmitters resembling satellites (pseudosatellites or pseudolites) is that their location is perfectly known, and from that their PL can be calculated in a stand-alone form. Also, if such systems are spread around the globe, a pseudo-GNSS can be created where satellite signals are difficult to acquire. Because ground-based transmitters use ground signals, data rates can also be increased, and therefore PL updating is increased.

When using a radio map created from actual measurements of the pseudolites' signal power, some of the problems with receiving satellite signals can be overcome, including the multipath problem, since trajectories can be predicted for users with low dynamics (even airplanes when landing).

What was a research question 20 years ago is a reality today: the use of a GNSS receiver together with a communications system such as the cellular telephone network. Improved positioning precision is still sought by using different techniques such as coherent signaling; multiantenna systems for space–time coding, also known as OFDMA or MC-CDMA; and so on. Differential positioning implementations are also being explored.

REFERENCES

[1] T. Ryan, Modern Experimental Design, John Wiley & Sons, 2007.

[2] B. Hofmann-Wellenhof et al., GNSS Global Navigation Satellite Systems: GPS, GLONASS, Galileo & More, Springer Wien, 2008.

[3] G. Beutler, Methods of Celestial Mechanics. vol. 1: Physical, Mathematical, and Numerical Principles, Springer-Verlag, 2005.

[4] S. Grewal et al., Global Positioning Systems, Inertial Navigation and Integration, John Wiley & Sons, 2001.

[5] D.D. McCarthy, International Rotation Service. IERS Conventions, 1996, p. 95.

[6] T. Kuhn, The Structure of Scientific Revolutions 2nd edition, University, of Chicago Press, 1970.

[7] N. Samama, Global Positioning: Technologies and Performance, John Wiley & Sons, 2008.

[8] G. Xu, GPS Theory, Algorithms and Applications, Springer-Verlag, 2003.

[9] Y. Bar-Shalom et al., Estimation with Applications to Tracking and Navigation: Theory, Algorithms, and Software, John Wiley & Sons, 2001.

[10] A. Leick, GPS Satellite Surveying, 3rd edition, John Wiley & Sons, 2004.

[11] R. Rogers, Applied Mathematics in Integrated Navigation Systems, 2nd edition, AIAA Education Series, 2003.

[12] GPS World. Available at *www.gpsworld.com.*

2D	Two-dimensional space
2G	Second-generation cellular technology
3G	Third-generation cellular technology
3GPP	3G Partnership Project
3GPP2	3G Partnership Project 2
4G	Fourth-generation wireless technology
ACN	Automatic crash notification
ALI	Automatic location information
AMPS	Advanced mobile phone systems
AOA	Angle of arrival
AP	Access point
APIT	Approximate point in triangulation
APS	Ad hoc positioning system
ARQ	Automatic repeat request
B3G	Broadband 3G
BCCH	Broadcast control channel
BER	Bit error rate
BS	Base station
BW	Bandwidth
CDMA	Code division multiple access
CEP	Circular error probability
CIR	Carrier-to-interference ratio
COG	Center of gravity
CRB	Cramer-Rao bound
DAB	Digital audio broadcasting
DoD	Department of Defense
DR	Dead-reckoning
DRE	Direct routing environment
DRP	Dead-reckoning path

DV	Distance vector
DVB	Digital video broadcasting
EDGE	Enhanced data rate for global evolution
EGPRS	Enhanced GPRS
ESPRIT	Estimation of signal parameters via rotational invariance techniques
ETSI	European Telecommunications Standards Institute
FCC	Federal Communications Commission
FDMA	Frequency division multiple access
GC	Generalized correlation
GDOP	Geometric dilution of precision
GPS	Global positioning system
GPRS	General packet radio service
GSM	Global system for mobile communications
HLR	Home location register
HSCSD	High-speed circuit-switched data
HSDPA	High-speed downlink packet access
ISI	Intersymbol interference
LA-ID	Location area identifier
LBS	Location-based services
LC	Location center
LMU	Location mobile unit
LOS	Line of sight
LS	Least squares
LSC	Location service center
LTE	Long-term evolution
LTS	Locating television systems
MAI	Multiple access interference
MANET	Mobile ad hoc network
MDS	Multidimensional scaling
MIMO	Multiple-input multiple-output
MISO	Multiple-input single-output
ML	Maximum likelihood
MLSE	Maximum likelihood sequence estimation
MMSE	Minimum mean square error
MSC	Mobile switching center
MUSIC	Multiple signal classification
NMTS	Nordic Mobile Telephone Service

OFDM	Orthogonal frequency-division multiplexing
OSI	Open-system interconnection
OTD	Observed time difference
PCS	Personal communications systems
PDA	Personal digital assistant
PHAT	Phase transform
PIT	Point-in-triangulation test
PL	Position location
PLI	Position location information
PN	Pseudonoise
POTS	Plain old telephone service
PSTN	Public switched telephone network
RF	Radio frequency
RSS	Received signal strength
SDP	Super-resolution delay profile
SDR	Software-defined radio
SIMO	Single input–multiple output
SISO	Single input–single output
SNR	Signal-to-noise ratio
SRE	Spread routing environment
TA	Time advance
TCH	Traffic channel
TDMA	Time division multiple access
TDOA	Time difference of arrival
TLS	Total least squares
TOA	Time of arrival
TRE	Typical routing environment
TSOA	Time sum of arrival
UHF	Ultra-high frequency
UMTS	Universal mobile telecommunication system
UTM	Universal Transverse Mercator
UWB	Ultra wideband
VHF	Very high frequency
VLR	Visitor location register
WCDMA	Wideband CDMA
WiMax	Worldwide interoperability for microwave access
WLAN	Wireless local area network
WML	Wireless Markup Language

A

Note: Page numbers followed by "f" denote figures; those followed by "t" denote tables.

265

Printed in the United States
By Bookmasters